Anthony C. Webster

Technological Advance in Japanese Building Design and Construction

Anthony C. Webster

Columbia University
Graduate School of Architecture
Planning and Preservation

Technological Advance in Japanese Building Design and Construction

Published by
ASCE Press
American Society of Civil Engineers
345 East 47th Street
New York, New York 10017-2398

Abstract

The book provides an overview of the technological capabilities of the Japanese building design and construction industry. The influence on Japan's industry exerted by the country's largest design and construction firms is described. Using case studies, the striking technological capabilities of these comprehensive construction companies are outlined, as used in building design and construction and in research and development. The industry's relationships with Japan's government and universities are also explored. The book also assesses the effect of the recent economic downturn of the Japanese real estate industry, the implications of contemporaneous construction-industry scandals, and the competitiveness of the United States industry relative to Japan's. In addition, tables summarize the operations, size and R&D work of Japan's largest building design and construction firms, while approximately 65 illustrations augment and amplify the text.

Library of Congress
Cataloging-in-Publication Data

Webster, Anthony C.
Technological advance in Japanese building design and construction/Anthony C. Webster.
p.   cm.
Includes index.
ISBN 0-87262-932-5
1. Construction industry – Japan – Technological innovations. I. Title.
HD9715.J22W4 1994   94-50713
338.4'569'00952—dc20   CIP

Library of Congress Catalog Card No: 94-50713
ISBN 0-87262-932-5
Manufactured in the United States of America.

Editing:
Ann Kaufman Webster

Copyediting:
Louise Rosa

Research Assistance:
Sergio Duran, Kimberly Holden, Erwin Viray, Jon Weiss, Dan Wood

Design:
Willi Kunz Associates Inc.

Contents

	Foreword	
1	Introduction	11
2	Technology Today: Building Design and Construction	17
3	Technology Tomorrow: Research and Development	55
4	The Role of Government and Higher Education	83
5	Technology and the Competitive Edge	95
	Index	103

This is the first of three books which focus on research performed by members of Columbia University's Graduate School of Architecture on the history and contemporary trends of the Japanese construction industry. The ongoing research is sponsored by Japan's Building Contractors Society. The other forthcoming titles in the series include Origins and Evolution of the Japanese Building Design and Construction Industry, *and* Japanese Building Design and Construction: an Overview.

In this book, Professor Anthony Webster provides a summary of the technological capabilities of the Japanese building design and construction industry including the influence on Japan's industry exerted by the largest, most influential firms. The striking technological capabilities of these "comprehensive construction companies" are outlined using exemplary case studies. At times expressing his personal opinion, Professor Webster also discusses and questions the comprehensives' building design and construction methods, while analyzing in depth their research and development (R&D) efforts. The industry's relationships to Japan's government and universities and aspects of Japan's industry are compared to those of the United States. The book assesses the effect of recent macro-economic changes to the Japanese economy, the implications of contemporaneous construction-industry scandals, and the competitiveness of the United States industry relative to Japan in the global marketplace.

The forthcoming book Origins and Evolution of the Japanese Building Design and Construction Industry *is divided into two major sections:* Traditional Japanese Building Practice, *and* Japanese Architecture, Technology and Construction in the Meiji Period, *written by Kenneth Frampton and Kunio Kudo, respectively. The* Traditional Japanese Building Practice *section focuses on the Japanese architectural system from its roots in Chinese Buddhism, to its status in 1868. The rise to preeminence of the "Master Carpenter" families is described, as well as the development of the* kiwari *and* tatami *modular systems and the standardization they brought to domestic construction under Shogunite rule. The book also explores linkage between style, materials and quality of construction. The section entitled* Japanese Architecture, Technology and Construction in the Meiji Period *details the influence of Western construction methods on the transition of Japan's construction practice during this period, and describes the resulting assimilation and mastery of Western methods with the establishment of Japan's modern construction industry and architectural profession.*

Japanese Building Design and Construction: an Overview *will provide a broad survey of the comprehensive construction companies, primarily through case studies of contemporary projects. The main areas of focus will include a description of the scope of their typical building projects and the methods used by these firms to execute them, including a tight-knit team approach across technical and design disciplines, and the design-build system. The book will also outline the uniquely Japanese cultural influences on the comprehensive constructors' operations, their long term client relationships, land values, neighborhood approvals of new projects, contractual arrangements, and corporate employment practices.*

On behalf of all of Columbia's staff involved in this project, I would like to express my deep appreciation to the Building Contractors Society and thank them for their support.

Bernard Tschumi,
Dean

Acknowledgements

This book was made possible by the generous support of Japan's Building Contractors Society and Columbia University's Graduate School of Architecture, Planning and Preservation. The support of officials from Japan's Ministry of Construction and Tokyo University have also been instrumental in completing the work. Many colleagues have provided invaluable assistance and support, for which I am grateful. I would like to thank here those individuals whose contributions were particularly significant: Professor Bud Griffis of Columbia University, Dr. Marc Weiss of the Office of Housing and Urban Development, and Harvey Bernstein, president of the Civil Engineering Research Foundation for their encouragement; the ASCE's anonymous reviewers, whose critical reading and comments were extremely helpful; and Nina Kramer, the ASCE's manager of journals, books and acquisitions, whose timely responses and diligence made publication of the book by the ASCE possible. Finally, I would like to thank Dean Bernard Tschumi for his support, Professor Kunio Kudo for arranging a productive research trip to Japan, and Professor Ken Kaplan of Harvard University for his insights.

Chapter 1

Introduction

Both Japan's urban landscapes and the engineering details of its buildings differ somewhat from structures built in American cities. Japan's high-rise commercial buildings, for example, are usually shorter than comparable buildings in the United States, due to the ever-present earthquake threat in the Japanese archipelago. For the same reason, the beams and columns comprising the skeletons of Japan's buildings are typically more substantial than the structural frames in the U.S. These differences aside, Japan's urban buildings bear a clear resemblance to those seen in American cities. The design and construction technologies used to make buildings in Japan are also usually recognizable variations or refinements of their American equivalents. Except in a few areas, such as construction automation and hybrid structural systems, the industries in both countries are following similar evolutionary paths, although the level of evolution varies, and each country is more sophisticated than the other in some areas.[1] Exotic structural systems and construction materials are almost as rare in Tokyo as in New York; the materials and methods employed in Japan are for the most part very similar to those used in the U.S. (figure 1).

Although the products of the Japanese and American building construction industries are similar, the types of companies that design and erect each country's major urban buildings, as well as the companies' relationships to their national governments and higher education systems, are strikingly disparate. The two countries' building industries are also set apart by the commitment of Japan's largest construction firms and national government to research and development.[2]

Prominent large-scale, commercial and residential urban structures are usually designed in the U.S. by architectural and engineering firms that specialize in building commissions. These structures are more often than not built by construction firms that similarly limit their scope of operations. Although large "design-build" and "engineer-constructor" firms like the M. W. Kellogg and Bechtel Corporations have added to America's urban landscape, their contributions to commercial and residential building are modest (unlike their dominant role in the creation of industrial structures).[3] In Japan, by contrast, the country's largest design-build construction firms have made a significant contribution to the commercial and residential architecture of Japan's cities, and they have engineered and constructed many buildings designed by independent architects. Works like Tokyo's Sendagaya Intes Building (figure 2), designed and constructed by the Takenaka Corporation (the fourth largest Japanese design-build construction firm) take their place alongside buildings like architect Fumio Maki's Tepia Center among the city's structural icons. Kenzo Tange's world-renowned National Gymnasium in Tokyo, constructed and partially engineered by the Shimizu Corporation (Japan's largest design-build firm) is emblematic of the role these firms play in constructing the designs of Japan's leading architects.

Shimizu and Takenaka are part of a group of Japan's six largest construction firms (also including the Obayashi, Taisei, Kajima, and Kumagai-Gumi Corporations) called the "Big Six." The Big Six corporations enjoy significantly higher sales volumes than Japanese companies with the next largest incomes (see Table I, chapter 3).[4] Except for the Takenaka Corporation, the Big Six firms – like most of Japan's large construction companies – design and build major bridges, dams, and other civil engineering works in addition to conspicuous architectural projects. The Big Six companies control roughly 10 percent of Japan's total design and construction market, including both building and civil engineering work.[5] These companies are similar in size to the largest construction-related firms in the U.S. In 1991, the Bechtel and Fluor Daniel Corporations were the only comparable American firms to record larger sales volumes than the Shimizu Corporation.[6]

Of the approximately 520,000 construction-related firms in Japan (about the same as the number of equivalent firms in the U.S.), the largest 2,900 are approved by the Ministry of Construction as "Specialized License Contractors" (SLCs), a classification required for bidding on many of Japan's largest civil engineering and building commissions.[7] To qualify as an SLC a firm must, among other things, meet certain size requirements and demonstrate its ability to design and construct both architectural and civil engineering projects. Approximately 1,000 of the largest SLCs are capitalized at more than 1 billion yen and control 27 percent of the industry's contracts. These firms are often referred to as "Zenecons," or "general contractors" by the Japanese. The Zenecons aggressively pursue design-build contracts, and prefer these commissions to other methods of design and construction.

The Zenecons' pervasive role in urban architectural design and construction distinguishes them from the smaller, more specialized companies performing the same functions in the U.S. In designing and building high-end architectural commissions, the Zenecons single-handedly perform the roles of America's specialized architectural, engineering, and building-construction firms.[8]

In addition to their architectural and engineering design-construction operations, the Big Six firms also have substantial research and development (R&D) divisions, as do the next largest twenty-four or so

1

2

1. Installation of curtain-wall at Sendagaya Intes Building. Takenaka Corporation. Tokyo, 1991.

2. Sendagaya Intes Building. General View. Takenaka Corporation. Tokyo, 1992.

3. Obayashi Corporation Technical Research Institute. Tokyo.
Land area: 71,486m^2
Building area: 11,175m^2
Total floor area: 18,006m^2
A. Office Building (Super Energy Conservation Building)
B. Geotechnics, Chemistry and Materials Laboratories
C. Acoustics Environmental Engineering Laboratories
D. Clean Room
E. Coal Storage Silo Testing Laboratory
F. Air Dome Test Prototype
G. Large Scale Geotechnical Engineering Laboratory
H. Multi-use Laboratory
I. High Technology R & D Center (Base Isolation Building)
J. Concrete Laboratory
K. Wind Tunnel
L. Vibration/Shaking Table
M. Large Scale Structural Engineering Laboratory

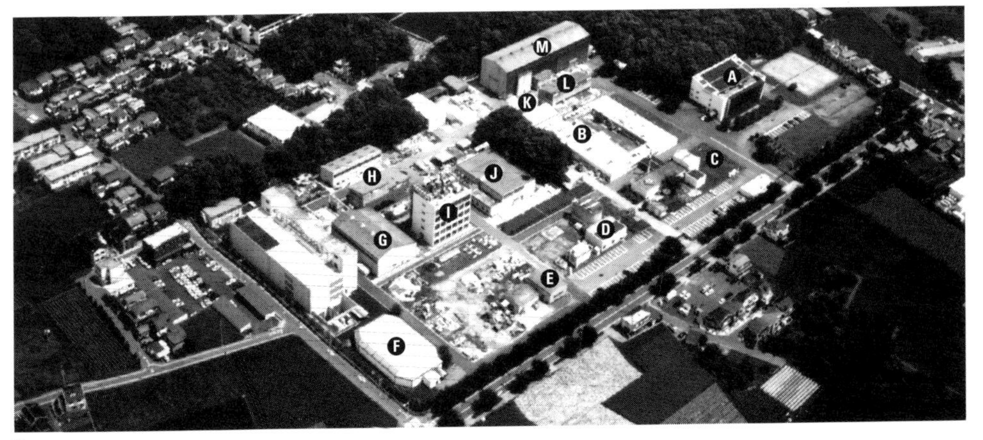

3

Zenecons. The R&D operations of the Big Six firms are particularly impressive. Their research facilities are typically "larger and better equipped than the largest university-based labs in the U.S. or elsewhere"[9] (figure 3); their funding averages almost 1 percent of sales, compared to America's 0.1 percent for firms of similar size;[10] their staff is mostly drawn from graduates holding Ph.D. or master's degrees from Japan's best universities; and their work is generally of a high quality as measured by, for example, the frequency with which their research is published in prestigious American engineering journals. R&D work performed by these Zenecons includes project support for design and construction, developing new building products, and testing prototype building systems.[11] Although most R&D is geared toward quickly marketable products or techniques, a significant number of projects have timelines of ten years or more. This commitment to protracted research reflects the corporations' emphasis overall on long-term planning over short-term market gains.

These thirty firms also provide some combination of planning, maintenance and financial services for big commissions. They obtain building design and construction contracts worldwide as well (the Big Six companies all have offices in the U.S.). Not infrequently, they become part owners of these projects.[12] The R&D activities and broad day-to-day operations of the largest Zenecons generally have no counterpart within a single firm in the U.S., and separate these companies from their smaller Japanese rivals. To distinguish these highly diversified firms, they are referred to as "comprehensive construction companies," or simply "comprehensives" in this book.

Unlike American firms, the comprehensive construction companies work fairly closely together – often on joint ventures – and have similar agendas for corporate growth. They all espouse three goals as crucial cornerstones to their long-term success: high quality, contained costs and project completion. The similarities among the firms make them cohesive, and help make them the most important force in shaping the future of Japan's construction industry.

Differences between the Japanese and American building design and construction industries are not due only to the different types of companies leading the industry in each country. Variations in academic training, government influence and culture also help to distinguish each nation's industry.

Since the Meiji era, Japan's university system has emphasized engineering prowess over architectural design skills. Today, at Japan's best universities, all students majoring in any building-design and construction discipline are trained within the architecture department of an engineering faculty.[13] These analytically oriented departments supply the comprehensives with a steady stream of technically proficient engineers *and* architects. Although the R&D capabilities of Japan's prominent architecture departments are significant, they are dwarfed in funding and equipment by the comprehensives' R&D divisions, which often lead Japan's construction research efforts.

Japan's government, like America's, regulates its construction industry. But the Japanese government also helps its industry develop technologically. The Ministry of Construction, which oversees the construction industry, encourages the creation of new technologies, particularly for automated construction. Though the funding levels for the Ministry's Building Research Institute are not as large as the top comprehensives, it initiates collaborative R&D efforts between government and industry on an annual basis. The Building Research Institute also pursues its own independent research, and collaborative projects with universities. Like the comprehensives' individual R&D efforts, many of the Building Research Institute's initiatives are long term projects, with no marketable results expected for a decade or more.

Many cultural factors, such as the Japanese emphasis on team effort over individual performance, help shape the unique face of Japan's construction industry. Possibly the most important factor is the pervasive belief in technological solutions to a wide range of problems – both technical and social. This notion encourages the continual development of new technologies for building design, construction and operation; it also encourages the comprehensives, the government and universities to strive for technical solutions to problems as diverse as labor shortages and the thermal performance of building facades.

Both the recent burst of Japan's bubble economy and the country's seemingly endemic construction-related bribery scandals have taken their toll on its building design and construction industry. Still, the deleterious effect of these problems is less than one might expect. Bribery in Japan's construction industry, unlike America's, has figured most often in contract procurement, not the in acceptance of shoddy work. Competition in quality and technology remains strong in Japan. In addition, Japan's construction economics is cyclical, and the comprehensives remain healthy enough to respond quickly to the next upturn. Even now (1993), the comprehensives are continuing to improve and broaden their technological capabilities faster than most American firms.

Japan's higher education system, its belief in technology's ability to help solve a wide range of problems, the legitimate relationships between its government, universities, and its construction industry, and the industry's demographics and long-term focus, all continue to fuel the comprehensives' technological development. When the international real-estate markets recover, these factors may help the comprehensives leapfrog the technical capabilities of America's leading building design and construction firms. Concomitant increases in efficiency could dramatically increase the financial success of these firms in Japanese and global markets.

Chapter 1 Endnotes

1. For a general overview of Japan's construction industry, see: J. Bennett, R. Flanagan, and G. Norman, *Capitol and Counties Report: Japanese Construction Industry* (Center for Strategic Studies in Construction, University of Reading, United Kingdom, 1987); F. Hasegawa, *Built By Japan* (New York: John Wiley and Sons, 1988); S. Levy, *Japanese Construction: An American Perspective* (New York: Van Nostrand, 1990).
2. Civil Engineering Research Foundation, *Transferring Research Into Practice: Lessons from Japan's Construction Industry* (Washington, D. C., 1991).
3. "The Top 400 Contractors," *Engineering News Record* (May 24, 1993).
4. For a detailed description of the Big Six, see: Sidney Levy, *Japan's Big Six* (New York: McGraw Hill, 1993).
5. Hasegawa, *Built By Japan.*
6. "The Top 400 Contractors," *Engineering News Record* (May 21, 1992).
7. Author's correspondence with Japan's Building Contractors Society and the Shimizu Corporation, Tokyo.
8. A. Webster, "Japanese Building Design and Construction Technologies," *Journal of Issues in Engineering Education and Practice* (October 1993).
9. B. Paulson, *Japan Technology and Evaluation Center (JTEC) Panel Report on Construction Technologies in Japan,* NTIS Report no. PB91–100057 (June 1991), 20.
10. R. Tucker, et al., ed. *JTEC Panel Report* (June 1991).
11. A. Webster, "Japan's Zenecons," *Architectural Record* (December 1992).
12. Civil Engineering Research Foundation, *Transferring Research Into Practice: Lessons from Japan's Construction Industry,* chap. 7.
13. Japanese universities, like those in Britain, are organized into "faculties" that are equivalent to American "schools." Tokyo University's engineering faculty, for example, is analogous to Columbia University's School of Engineering.

Chapter 2

Technology Today: Building Design and Construction

Comprehensive Construction and Japan's Design-Build System

The comprehensives prefer building commissions for complete design and construction services, and the number of contracts they receive for these services is increasing. The comprehensives like design-build contracts, not just because of their large scope of work but also because of the non-adversarial relationships between designer and contractor inherent in this system. Design-build projects are executed by the comprehensives using in-house teams that include all the professionals required for both design and construction. By having designers and builders work together throughout a project, the comprehensives assert that jobs can be completed more quickly and economically, problems of buildability are often foreseen and resolved before construction begins, and quality control is more easily assured.[1] Because the design-build system demonstrates some potential advantages over America's more prevalent bid system, and because these commissions best illustrate the overall technological capabilities of the comprehensives, the system is worth examining in some detail.

The Design Team: Its Charter, Tools and Methods

Upon obtaining a typical, large-scale design-build commission, the comprehensive construction company assembles a senior-level team to run the job. The team typically includes architects, structural, mechanical, and electrical engineers, an estimator and engineers representing the firm's construction division. For many projects, a senior member of the company's R&D division also participates. Led by an architect, the team works closely together and meets formally about once a week throughout the design and construction of the project. Junior architects and engineers assigned to the job work in adjacent zones of a large, semi-open office floor (figure 1). On some projects, senior team-leaders move their workspace to the same area. The comprehensives sometimes accept design-build commissions in association with independent architects. In these cases the independent architect may not lead the team, and participates primarily in schematic design.[2] The comprehensives claim that their team approach fosters close and continuing communication among disciplines, and helps achieve the three goals they all consider crucial cornerstones of a project's success: on-time delivery, con-

1

2

1. Design Offices. Shimizu Corporation.

2. Mitsukoshi Department Store, Tokyo. Hazama Gumi, 1991.

tained costs and high quality. They point out that cost controls, for example, are to some extent built into the system by the team's makeup, which features more engineers than architects. The inclusion of construction engineers and cost estimators, who care deeply about efficient construction and the financial bottom line, tends to discourage the creation of spaces that take a long time to build or are expensive to construct.

The design-construction team is responsible for preparation of all of a project's design and construction documents, client contact and navigating the design through Japan's labyrinthine system of approvals. The construction of the fifteen-story (four-underground) Mitsukoshi Department Store, designed by Hazama Gumi with Yokokawa Architects of Tokyo, illustrates the typical process and results of the design-build system (figure 2). During the building's design and construction, Hazama created four major sets of plans and related documents, both for purposes of construction and for submission to three independent regulating bodies (figure 3). On the Mitsukoshi project, Hazama and Yokokawa worked together from October 1987 to March 1988 to create a set of "basic plans." After incorporating some client-inspired modifications, the two firms produced a set of "disaster prevention plans," describing the structure's fire and earthquake mitigation systems, which they submitted to local authorities in August 1988. Preparation of a set of "detailed documents," roughly equivalent to an American set of bid drawings and specifications for major architectural and engineering systems, was completed in January 1989. Portions of this set of documents were submitted to authorities to obtain construction approval. Finally, an exhaustive set of "comprehensive drawings," representing nearly all the components of the building (and therefore somewhat more exhaustive than American shop drawings) was begun in February 1989. The comprehensive drawings were used only to facilitate construction, and were not submitted to any government agencies. Construction of the work described in these drawings often began almost immediately after they were completed.

Both Hazama and Yokokawa had previously worked with Mitsukoshi. Such continuing relationships between client and builder are not uncommon among the comprehensives, and these affiliations have yielded some unusual methods of doing business. Hazama, for example, did not receive a contract (or any payment) for its work on the Mitsukoshi store until it submitted portions of the detailed documents to the government for construction approval. The comprehensives' established clients often pay for design services after they are rendered. This reimbursement process depends on both the relatively high levels of trust between Japanese owners and builders, and on a society that is significantly less litigious than the U.S. The process also underscores the comprehensives' emphasis on project cost, short construction time and material quality, rather than the subtleties of its architectural spaces. This de-emphasis on spatial design is not unique to the comprehensives; Japan's smaller Zenecons frequently market their architectural and engineering services "for free" when competing with their larger rivals. Through this strategy, they implicitly support the belief of many American developers – that the successful realization of a building project is more important than the quality of the spaces inside it.

Primary Building Systems

Structure

Because of the high level of seismicity in Japan, earthquake (and of course gravity) loads govern the structural design of most buildings. Although the structural systems used for earthquake resistance in Japanese construction are essentially the same as those employed in the U.S., more composite and hybrid structural components are used in Japan. To achieve a high level of ductility,[3] most medium and high-rise construction employs moment or eccentrically braced frames, which are made of reinforced concrete (RC), steel-reinforced concrete (SRC), or steel, depending on the building's height and intended use. SRC is a composite steel-reinforced concrete system employing either steel I-beams (wide-flange shapes), surrounded by reinforcing-bar and concrete, or tubular steel sections filled with concrete. Though precasting is often used in Japanese buildings, prestressing and post-tensioning operations are less common.

The cost of a medium or high-rise building in Japan can be as much as 40 percent more than its equivalent in the U.S., due mainly to Japan's high cost of materials.[4] Japan is a material-poor country with few local supplies of many building components. Despite the country's strength in steel manufacturing, most of Japan's iron ore is not obtained locally and its cheapest structural system for medium-rise buildings is often reinforced concrete framing. Concrete strengths are typically between 3 and 5 kips per square inch (ksi),[5] which usually makes structural frames of more than ten stories (designed for Japan's strict earthquake codes) too bulky unless they are constructed of SRC. Tall office buildings, with their increased mechanical requirements and longer spans often employ structural steel frames. High-rise residential structures are usually constructed of SRC because of its superior deflection characteristics under service loads and the relatively modest mechanical services required in these buildings.

A few different types of SRC are used, depending on architectural requirements, loads and size of the building. Typical uses of SRC are illustrated in two recently completed buildings constructed by the Mitsui Corporation at the River City 21 development in Tokyo. For the thirty-seven-story A Building (figure 4), which is partially subsidized

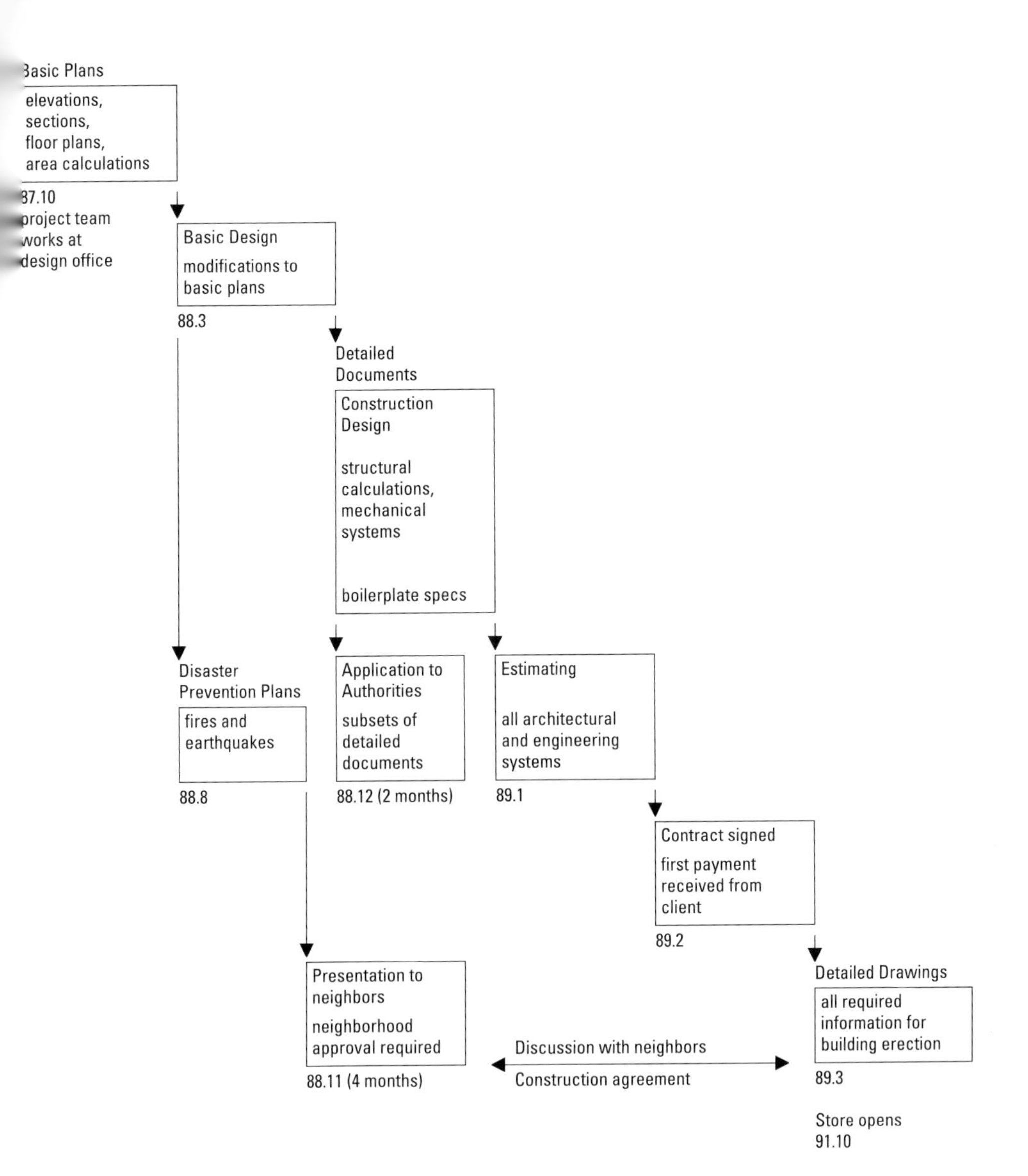

3. Mitsukoshi Department Store. Design and construction process.

housing, Mitsui made the structural frame of a proprietary system called Mitsui Steel-Concrete (MSC). The frame takes the form of two concentric tubes made of SRC moment frames, which support gravity loads and provide earthquake resistance. The columns are tubular steel, filled with concrete and covered with fireproofing after they are placed. Beams are wide flange shapes, encased in concrete with some light reinforcing. The encasement acts as fireproofing and provides a ledge for the placement of precast concrete planks. Concrete encasement is not continuous across columns, leaving moment continuity to be developed by the steel beams alone. A structural topping slab is poured over the concrete planks after installation to engage them as diaphragms.

The nearby, forty-story Building H, a luxury residence, uses a more common and more labor-intensive, nonproprietary SRC system. Here earthquake loads are resisted by a three-dimensional matrix of orthogonal moment frames. The columns are made of wide flange sections, with tees welded transversely to their webs, creating a built-up cruciform section. A reinforcing-bar cage is placed around the columns before the concrete is poured. Steel wide flange beams are caged with re-bar and encased in concrete, which is continuously connected across columns with steel re-bars. This system, with its re-bar encased, rolled column sections and the continuity of its reinforced concrete between beams and columns, is clearly more expensive than the MSC system, but it allows for a larger column grid with almost the same beam depths as those of Building A.

Tokyo's twenty-story Century Tower Office Building by Norman Foster, the Obayashi Corporation, and the Asahi Newspaper annex by Takenaka illustrate some typical methods of steel construction in Japan (figures 5 and 6). To resist earthquake and wind loads, Foster's building employs a two story eccentric braced frame system in one direction and moment frames in orthogonal planes. The Asahi annex, which is an addition to Asahi's headquarters building, employs moment frames in two directions. From a distance, both structures look similar to American systems. In addition to their increased lateral stiffening, the biggest differences between the structural systems of these buildings and comparable American ones are the profusion of field-welded connections, the locations of field splices, and the long column sections (sometimes three stories tall) which are installed in one piece (figure 7).

Vibration Control

Vibration control is an important part of Japanese building design because high land prices ensure the continued construction of tall, vibration prone buildings, and because Japanese customers are generally more sensitive to tall building movements than their U.S. counterparts. Japanese citizens' sensitivity to tall building vibrations is due in part to the omnipresent earthquake threat in Japan, and to their relative

4

5

4. River City 21. Building A.

5. Century Tower, Elevation. Norman Foster and Obayashi Corporation, Tokyo, 1991.

6. Asahi Newspaper Annex (Addition to Existing Building), Takenaka Corporation. Tokyo, 1992. (Photo by Taisuke Ogawa).

6

7

7. Asahi Newspaper Annex. Construction photo showing field-welded connections and long column sections.

8. Funabashi Dormitory, Takenaka Corporation, Tokyo, 1987.

8

unfamiliarity with skyscrapers. Yokohama's seventy-story Landmark Tower, which was engineered and constructed by a group of comprehensives led by Taisei, is currently Japan's tallest building at 296 meters (971 feet). This is shorter than a number of American skyscrapers, including the sixty-year-old Empire State Building, which weighs in at 102 stories and 320 meters (1050 feet). Japan has only about 800 buildings taller than 60 meters (200 feet). Since lateral building motions caused by both wind and earthquakes can often be reduced more cheaply with mechanical damping devices than by conventional structural stiffening systems, the motivation to develop mechanical vibration controls is very strong.

Both active and passive vibration control technologies have been employed in Japan with some success in recent years (Table I).[6] Passive technologies – including tuned mass, sloshing, and viscoelastic dampers, as well as base isolation systems – have been installed by the comprehensives in commercial structures[7] (figures 8 and 9) and have been the focus of much research in their R&D divisions. Active technologies, including active mass dampers and variable force tendons are currently under research, and in a few cases have been installed in for-profit structures. Although these systems will not substantially increase the strength of a building subject to a strong earthquake, their ability to mitigate the less powerful vibrations due to wind and small earthquakes, and their ability to increase the threshold of plastic deformations makes them an attractive way to improve a building's performance.

One of the most interesting Japanese "quasi-passive" vibration control systems is the pneumatically sprung floor designed by Kajima for the recently completed Edo-Tokyo Museum (figure 10). Kiyonori Kikutake is the museum's architect; Kajima is responsible for its construction and much of its engineering design. The bulk of the building's interior is suspended 65 meters above the ground by four super-columns. The main floor cantilevers 40 meters beyond the columns in the building's long elevation. During an earthquake, the building's unique shape causes the cantilevered floors to vibrate up and down due to horizontal ground motion (figure 11). To avert panic among museum visitors in such a scenario, Kajima has designed a double floor system that works much like an air-sprung truck suspension (figure 12). Air springs and viscous dampers separate the structural floor below from the finish floor above. The springs and dampers are tuned for optimum floor performance in the cantilever modes most highly excited during earthquakes. The system is called "quasi-passive" because level sensors and air pumps keep the floor level under the museum's highly variable static loads.

Because they are more efficient at reducing vibrations than passive and quasi-passive dampers, active mass dampers (AMDs) are currently the focus of much experimentation, both inside and outside Japan.

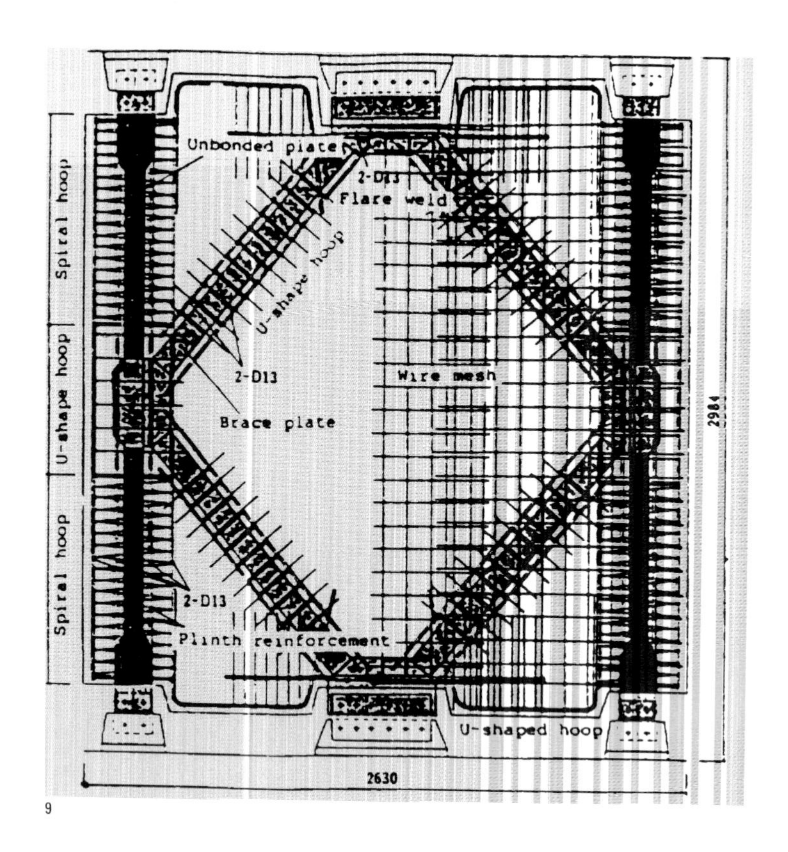
Unbonded plate
2-D13
Flare weld
U-shape hoop
2-D13
Wire mesh
Brace plate
2-D13
Plinth reinforcement
U-shaped hoop
Spiral hoop
U-shape hoop
Spiral hoop
2984
2630
9

10

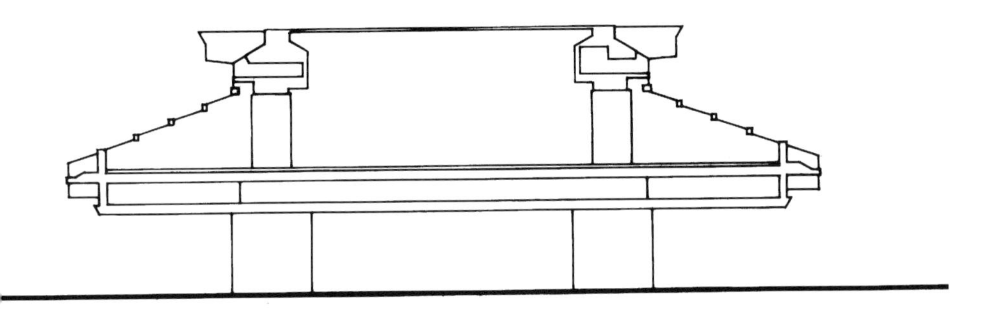

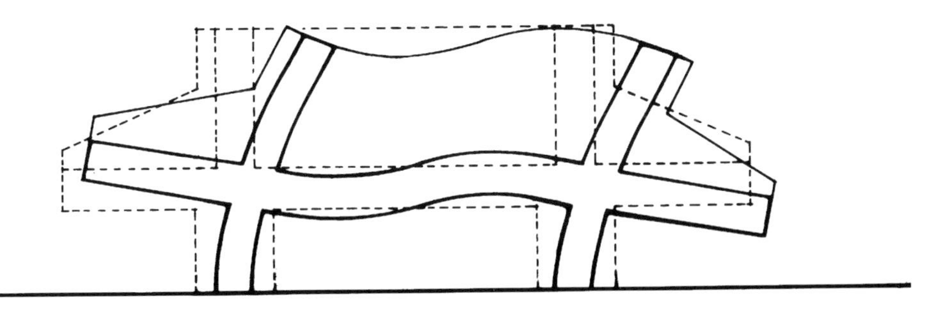

11

9. SROP wall system. Seavans South Building, Shimizu Corporation.

10. Edo-Tokyo Museum. General View.

11. Edo-Tokyo Museum. Schematic view of building motion during an earthquake.

12. Edo-Tokyo Museum. Double floor suspension system, detail.

12

AMDs are really tuned mass dampers (sharing with them such critical components as heavy masses, springs, and viscoelastic dashpots) that have been made more efficient by computer-controlled hydraulic jacks that optimize their movement.[8]

The Takenaka Corporation has recently installed an AMD system on its eleven-story Sendagaya Intes office building in Tokyo (figure 13 - the building is known in Japan as simply "Sendagaya Intes"). Its wing-like plan-geometry makes the building prone to large swaying and twisting motions during high winds or earthquakes. The central feature of the Intes's AMD system is a set of two 30-ton weights installed in its mechanical penthouse, which work to reduce both types of vibrations (figure 14). The weights are connected to the building by rubber springs and hydraulic jacks. As the building sways back and forth in strong winds or an earthquake, velocity sensors at the top, middle and base of the building transmit a record of its motion to a computer, which instructs the jacks to move the weights so as to minimize the building's motion.

Environmental Control

The Sendagaya Intes building also features an innovative environmental control system that works in concert with its structural vibration controls. While simple concrete and steel blocks are typically used for damper masses in the U.S. (the Citicorp Building's concrete tuned mass damper weights are a prominent example[9]), Takenaka made the unusual decision to construct the Intes Building's mass blocks of crystallized liquid ice tanks. Besides being a crucial part of the Intes's AMD system, the fluid in the tanks is frozen into an ice slurry at night, when utility rates are cheap. During the day, the ice is melted to help cool the building. By looking at the building's HVAC requirements together with its structural problems, Takenaka was able to overcome one of the biggest drawbacks of mass damper design – that of placing a large weight on top of a building – in a way that both improved the Intes Building's structural performance and increased its energy efficiency.

Hazama Gumi also used a cooled-fluid technique to improve the environmental performance of the Mitsukoshi Department Store. The sub-basement of the building is flooded with water that is cooled during the night and pumped to heat exchangers throughout the building during the day. Hazama reports that this system saves money by taking advantage of cheap nighttime electrical rates and reduces the size of the chillers needed to cool the building during peak shopping hours.

In addition to using these thermal buffering systems, the comprehensives have also explored ways to improve environmental performance by using individualized controls and "intelligent systems." The Fujita corporation has built one of the most ambitious individual climate control systems in use in Japan today. At its headquarters building in Tokyo, completed in 1992, individual thermostats and air ducts regulate

13

13. Sendagaya Intes Building. General View. (Photo by Kaneaki Monma).

14. Sendagaya Intes Building. AMD System.

14

15

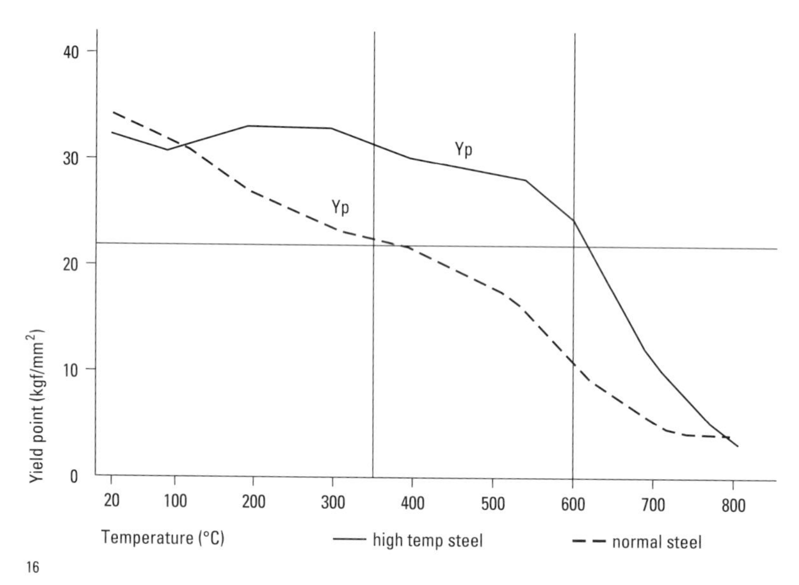

16

15. Fujita Headquarters building. Individual climate control system.

16. Temperature vs Yield Point Graph. Conventional vs fire-resistant steel.

the air temperature for each employee (figure 15); at a large conference room it built for the Onward Institute building in Yokohama City, cigarette smoke is pulled into a foul air stack by vacuum exhausts at each ash tray.

Takenaka's Intes Building provides a typical example of intelligent systems used for environmental control. Along the building facades, venetian blinds are opened and closed by a computer system that monitors incident sunlight conditions. The system optimizes overall energy performance and comfort by controlling the blinds to regulate the amount of sunlight and heat-inducing infrared radiation entering the building. By mechanically solving some of the building's heat gain problems, this system allowed the building's curtainwall to be made of a more transparent glass than would be otherwise acceptable. The transparent glass helped amplify the architect's intended expression of the building as a translucent wing.

Japan's steel companies are also looking at how a building's structure can play a more explicit role in its external appearance. Nippon Steel has recently developed a fire resistant, 50 ksi steel for building use, whose yield point does not significantly change at temperatures up to 500 degrees centigrade (figure 16).[10] Though it is still being tested, the use of this material in combination with intumescent paint should make it possible for exposed steel to have a two-hour fire rating. This will allow steel to be exposed in many more situations than is now safely possible. Depending on its cost, the new steel may also lower a building's overall expense by reducing the amount of hand-applied spray fireproofing required.

Building Enclosure

Some of the comprehensives are working on ways to reduce the weight of facade systems in order to minimize structural requirements and therefore costs. Premanufactured tiled facade panels, similar to those used on Fujita's headquarters building in Tokyo, are very popular in Japan (figure 17). These panel systems are often made of inexpensive but heavy reinforced concrete, significantly adding to a structure's dead load and necessitating a stocky, expensive structural system. Kenzo Tange's forty-eight-story Tokyo Metropolitan Government complex in Tokyo, for example, is clad in 6-inch-thick precast concrete panels faced with an inch of granite. Kajima corporation (among others) has recently developed tile-clad, carbon fiber reinforced concrete (CFRC) panels to reduce facade weights and thereby decrease the required strength and cost of a building's structural system (figure 18).

Glass and aluminum curtainwall facades, which are inherently lighter than most monolithic panel systems, are used less frequently in Japan than in the U.S. The "stick and glass" systems employed on the Intes Building and portions of the Century Tower are exceptions to common practice. The use of stick systems is limited in Japan, in part, by the relative premium for on-site labor that the comprehensives claim these systems require. Construction-site labor is scarce in Japan, so spending a bit more money on a heavy structure is often less costly than increasing field manpower requirements. Panelized facade systems are also favored because they are prefabricated in a factory, where quality is easier to control than at a construction site. Tile, an inexpensive facade-facing material, is popular in Japan, and easily lends itself to premanufactured panel construction. Structurally glazed curtain-walls, used with increasing frequency in the West, are not normally allowed in Japan. According to Sato Chikafusa, General Manager of Obayashi's Architecture and Engineering Division, Japan's Ministry of Construction (MOC) forbids the use of this technology for most applications because "the design loads and movements required for Japanese earthquakes are too large for today's structural silicones to accommodate."[11]

Disaster Prevention

Japan's disaster prevention codes, including structural provisions for earthquakes and general requirements for fire control, are significantly stricter than those in the U.S. The structural bulk and attendant costs required to meet earthquake codes contribute to the limited height of Japan's skyscrapers. The consequences of Japan's fire codes are seen more clearly in the spatial organization inside high-rise structures. Building codes require every floor to have its own fire exhaust system, and each level must be separated from adjacent floors with mechanical barriers. Atriums in high rise buildings are generally not permitted unless separated from each floor opening onto them by wire glass partitions and fire-shutters. Atrium spaces such as John Portman's Hyatt Regency Hotel in Atlanta, which relies primarily on incombustible construction and sprinklers to prevent fire spread, are inconceivable in Japan. One of the first large open atrium spaces constructed within Japan's strict constraints was Kajima's KI building in Tokyo, completed in 1989 (figure 19). The perimeter of the atrium features heat-controlled mechanical fire shutters (similar to those used in Louis Kahn's British Art Museum in New Haven) that automatically fall into place during a fire, separating each level from the adjacent floors and atrium space.

Design Technologies

The technologies used by the comprehensives for designing buildings are fairly similar to those employed in the U.S., as evidenced in their structural design techniques. Finite-element computer codes are used routinely by the comprehensives for structural analysis and design, as are computerized graphic representations of a building's structural behavior. These programs resemble those used by medium and large consulting engineering firms in the U.S.

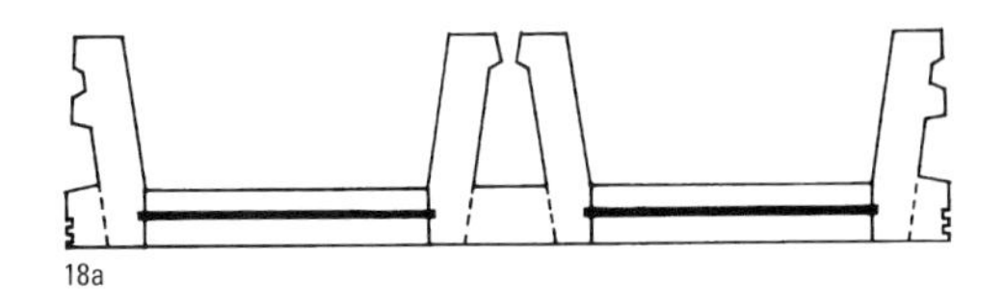

18a

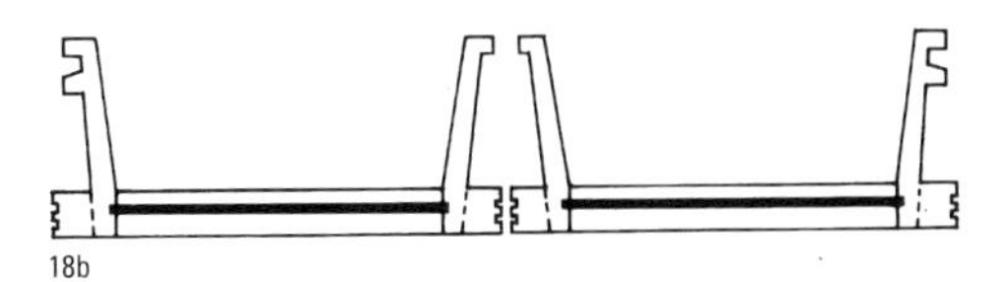

18b

17

17. Fujita Headquarters, Tokyo.

18a. Plan-section, conventionally reinforced panel. Weight: 5 tons.

18b. Plan-section, CFRC Facade Panel. Weight: 2 tons.

19. Kajima KI Building, Tokyo. Atrium.

19

Japan's design technologies do, however, differ significantly from American techniques in some areas. In their use of full-scale testing, computer simulation, and CAD integration, the comprehensives employ techniques that are little used or underdeveloped in the U.S.

Full-scale testing is a method used worldwide to verify the performance of architectural components before they are applied to entire buildings. Mockups of proposed curtainwall designs, for example, are routinely subjected to simulated hurricane-force wind and rain. Beyond testing facade systems, the comprehensives also test full-scale structural components and environmental control systems fairly regularly. Critical beam-column connections designed for Mitsui's H Building, for example, were tested for ductility before the structural steel for the entire building was ordered. Similarly, Fujita built and tested full-scale mockups of their individual environmental control and smoke exhaust systems for its headquarters and Onward Institute buildings, respectively.

These tests, which are usually performed by the comprehensives' R&D divisions, help ensure the quality of the project and sharpen the building designer's skills. Designers of the systems in question, as well as representatives of a project's construction team attend the tests, and sometimes observe the construction of the mockup assemblies. Besides ensuring that a building's crucial components will meet expected quality and performance standards, these tests confront architects and engineers directly with potential inefficiencies of their designs. They also help designers by exposing them to the techniques required to build what they have represented on paper. These tests are, of course, not without disadvantages: full scale prototypes take a long time to build and are costly to analyze. But the comprehensives view these as minor drawbacks. They feel that the cost of materials is in part recouped by savings inherent in the design-build system, and the time spent is compensated by the knowledge gained by designers, who presumably are able to create a wider variety of more efficient systems as a result of their experience with testing.

Computer simulation is used by the comprehensives in much the same way it is used in the U.S. However, the comprehensives also apply their extensive simulation techniques to solving problems that would require the assistance of university personnel or specialized consultants in America. Tokyo's clay soil and high seismicity make foundation design difficult in that city, especially for tall buildings. To design foundations for large commercial structures in the area, virtually all of the comprehensives have developed soil-structure interaction programs to model soil settlement and building performance during earthquakes. Kajima, for example, has written programs employing wave propagation theory and finite element methods to design the pile supported foundations of numerous highrise buildings, including The World Business

Garden building (1991) and The Manhattan hotel (1991) in Makuhari. Programs that perform similar analyses are available in the U.S., but their complexity (and until recently, their need for lots of computer power) makes them uncommon outside universities. In the West and in Japan, access to these programs will grow as the price of the powerful computers required to run them continues to drop. Similarly, the comprehensives use atmospheric-modeling programs to predict the environmental control performance of complicated buildings' HVAC systems, and to establish the performance of their fire control smoke exhaust systems. Shimizu's 1991 Seavans Headquarters Building in Tokyo, was designed using both types of computer systems (figure 20). Like the comprehensives' soil-structure interaction programs, these sophisticated types of computer codes are generally available in the U.S. only at universities and the most advanced consulting engineering firms.

At first glance, the CAD systems typically employed by the comprehensives do not seem very different from those used in the U.S. Production drawings are often created with two-dimensional CAD systems, while architects use three-dimensional systems for spatial design and client presentation. Three-dimensional working drawing CAD systems recently developed in this country by Columbia University's Bud Griffis, among others, currently have no counterpart in Japan.[12] Also, CAD-CAM methods, such as those used by Starnet Structures to make robotically machined structural components from the information in a CAD database, are not used by the comprehensives (although some architectural divisions have models machined directly from their CAD drawings). Similarly, CAD analysis-design packages, such as the composite-floor design and drafting package being developed in the U.S. by the Computers and Structures Corporation, are not in widespread use.

Although the comprehensives' CAD systems are not more advanced than American systems, the way the they use CAD to coordinate all of a building's design documents is more impressive. In some of their most technologically ambitious projects, the comprehensives link their structural, architectural, and mechanical drawings, which progress in parallel as the building's design is refined. Takenaka's design of the Asahi Newspaper annex provides a good example of the comprehensives' state-of-the-art techniques. During the design process, a CAD-based set of structural floor plans and sections, including column dimensions, slab widths and beam depths, was produced. This set became the electronic template from which architectural, mechanical and electrical drawings were derived. After the design of the building's structural system was essentially complete, construction of the building started, leaving many decisions of architectural materials, secondary mechanical ducting and electrical layout to be determined while the building was being constructed. Using the base structural, architectural and mechani-

20

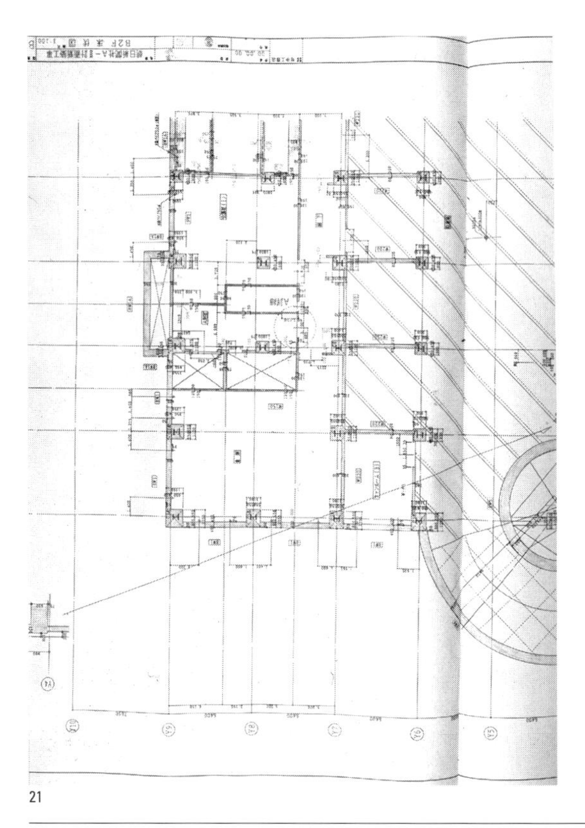

21

20. Seavans South. Shimizu Corporation, Tokyo, 1991.

21. Asahi Annex. Typical Structural Base Drawing.

cal design drawings, CAD-based construction documents for the building's partitions, finishes, electrical wiring and HVAC outlets were produced during this time (figures 21 and 22).

This system of consistent electronic templates helped expedite the design and construction of the building in two ways: first, because the design team was working from the same base drawing set, errors in the design of the building's major architectural, structural and HVAC systems were reduced. Second, by working directly from scaled-up versions of base architectural and structural drawings, those charged with detailing the building while it was being constructed could work faster and with more confidence than they would have if they had worked by hand, looking up base dimensions from scratch.

As impressive as this method is, it could be easily duplicated by more consistent use of CAD among consultants in the U.S. This fact underlines one of the advantages of Japan's design-build system and the comprehensives' methods. By controlling all aspects of a building's design and construction, it is easy to dictate the technologies used to design its components, and, by integrating them, to streamline the design process. This type of integration is almost impossible when many individual consultants and contractors share responsibility for various portions of a structure's design and construction. The comprehensives' design-build methods also have a significant impact on the type of structures they create. Unlike architects in the U.S., who rule over their technical consultants and largely control a building's design (if not its construction), the comprehensives' architects are equal partners on a team primarily comprised of technically oriented members. These members often care more for efficiency, economy and quality of workmanship than spatial elegance, so the team's collaborative results are often architecturally disappointing.

The Construction Team and its Techniques

Team Membership, Location and Methods

The Japanese building construction site is not just the place where a structure is erected but also where the construction process is managed and final designs are completed. Many tasks that are usually tackled at home offices in the U.S., including construction planning, drafting of many details and checking of shop drawings, are performed at the building site. The construction site operations at Takenaka's Asahi Newspaper annex (figure 23) illustrate how a state-of-the-art construction office operates.

At the onset of construction, Takenaka did not move a series of trailers to the job site for their field staff. Instead, they erected a prefab-

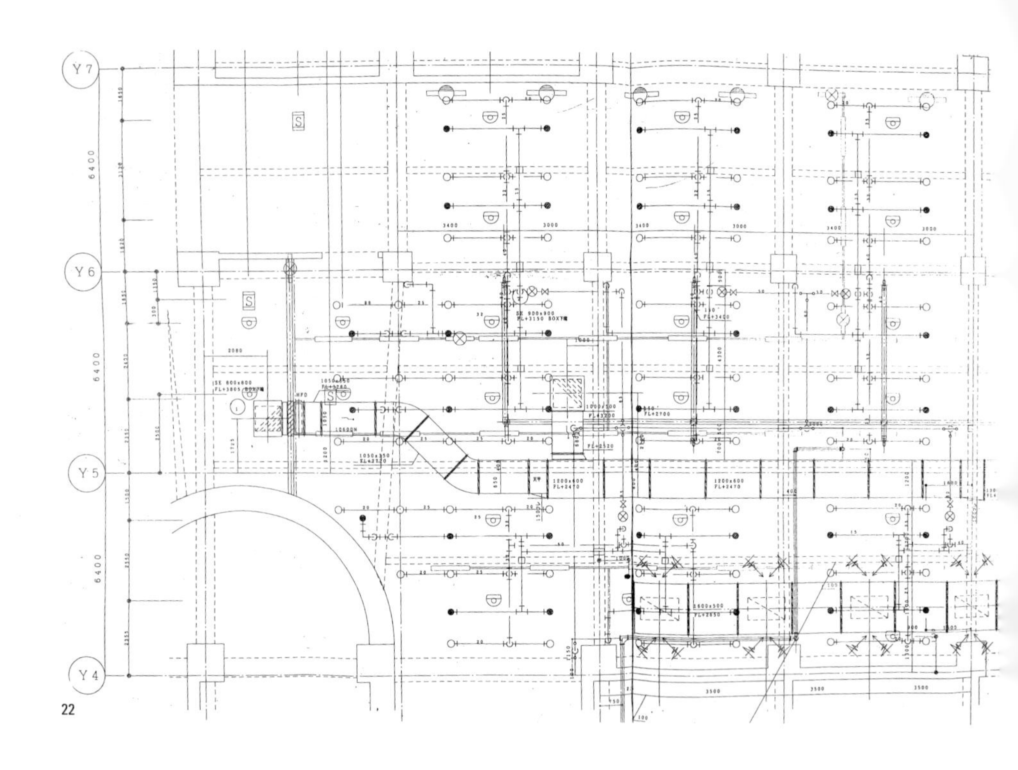

22

23

22. Asahi Annex. Detail drawing showing typical structural drawing used a an electronic underlay.

23. Asahi Annex. Construction Field office wrapped among portions of the original building. Takenaka Corporation.

ricated, two-story building on stilts over a pedestrian walkway, complete with its own HVAC and plumbing systems (figure 24). Working inside were structural, mechanical, electrical and construction engineers, a few architects and about eight CAD operators. Equipment in the building, besides the expected fax machines and personal computers, included nine CAD workstations and a full-size color plotter. Documents produced by the field office staff while construction proceeded included most of the building's finish schedules and details, complex ductwork drawings and electrical system details.[13] The on-site staff was also responsible for coordinating and checking all of the building's steel and re-bar shop drawings, as well as performing all required construction planning and supervision. Almost all of this work proceeded in step with the building's construction. The construction site staff also ordered the shipment of many non-critical materials immediately prior to their installation in the building.

Takenaka's construction scheme draws on paradigms from both the automotive and electronics industries. During the 1970s and early 1980s, Japan's auto industry pioneered the "just-in-time" manufacturing approach, delivering components to its assembly lines at the last possible minute.[14] This saved money on inventory, and, by making component substitutions easier, created more flexible assembly processes. As Tracy Kidder describes in *The Soul of a New Machine*,[15] America's minicomputers of the same period were *designed* using analogous methods. The overall logic of a proposed central processing unit (CPU) was laid out first, and the resolution of many design details was put off as long as possible. Many of the computer's logical subsystems were relegated to individual but complex chips. In this way, the overall circuitry of a new computer was completely laid out before the details of many of its chips were prescribed. The tedious design of these chips was postponed until just before the computer was prototyped; this allowed fabrication of the main circuit boards to begin as the design of their chips was being finalized.

By extending some aspects of design into the construction phase, and by dealing with particular materials and assembly processes as they are needed, Takenaka mimics these auto and computer industry techniques to maximize the efficiency of its design/construction process. As Takenaka aptly describes its methods, it "encourages timely decisions and recognizes that they must often take place at the site during construction."

Borrowing other industries' methods for their design and construction operations (as exemplified by Takenaka at Asahi) is commonplace among the comprehensives. Japan's modernization during the Meiji era was based in large part on the transfer of Western technologies to Japan. Today, the Meiji era's corporate descendants maintain a healthy unselfconsciousness about adopting good ideas and techniques wherever they find them.

Mitsui also benefited from a just-in-time approach to the purchase of the facade systems for its River City buildings. Mitsui's field engineers did not decide on facade materials for its River City Building I until after the construction of the structural frame had begun. As construction of Building I's structure was underway, Mitsui's construction team reviewed the cost and performance of facade systems it had used on recently completed jobs, and decided to use a newly developed, lighter and cheaper system for the new building. By allowing its construction engineers to decide on facade systems at the last minute, Mitsui freed them to use the latest and most inexpensive technology, while saving money by postponing purchases as long as possible.

Construction Technology

One of the biggest challenges facing Japan's construction industry is finding enough laborers to build its buildings. Construction work – from excavating for foundations to applying building finishes – has the reputation of being difficult (*kitsui*), dirty (*kitanai*) and dangerous (*kiken*). These three perceived characteristics, known commonly as the three K's in Japanese, have driven much of Japan's semiskilled workforce to safer, "cleaner" jobs in the manufacturing and service industries. Compounding the labor problem is Japan's education system, which produces young adults whose levels of literacy and mathematical proficiency naturally guide them toward more highly skilled positions.[16] The numerous recent bribery scandals involving the building construction industry have also tarnished its image among potential blue and white-collar employees.

The low esteem in which construction work is held is demonstrated by the economic and social class of the industry's laborers. Ninety percent of the adult male workforce in Osaka's working class Arin district, for example, is employed as casual labor in the construction industry. The area is so poor that it has been described as "the nearest thing in Japan to a smoldering ghetto,"[17] and its inhabitants are so discontented that riot police have been called on more than once to keep the streets calm.

The comprehensives are responding to these problems in several ways. Most obvious at construction sites is the effort to make the workplace cleaner and safer. The comprehensives' job sites are much less messy and chaotic than comparable American workplaces, and concern for worker safety is evident everywhere. Safety is, along with quality, cost and time of construction, one of the four stated primary management concerns at the comprehensives' job sites. Field employees are assigned to safety management in the same manner that others are assigned to, for example, cost control. The most visible symbols of the safety man-

24

24. Asahi Annex, Takenaka Field Office. Interior view.

25. Safety wrap around the Asahi Annex (in construction). Takenaka Corporation, Tokyo, 1991.

25

agers are the prominent cocoons of plastic mesh that sheath nearly all of Japan's tall building construction projects; the mesh keeps workers from slipping off the building, and protects passersby from falling tools and debris (figure 25). Safety posters, mounted throughout a building construction site and concentrated near worker entrances also help keep safety consciousness high (figure 26). Government actions also encourage safety: in 1990, Kumagai Gumi was prohibited from bidding on public works for six months because of an accident at one of its job sites.[18]

The comprehensives are also responding to the problems of worker safety and discontent with a variety of technologies that either streamline the construction process or reduce the amount of manual labor required in the field. Structural and nonstructural components are prefabricated whenever possible into subassemblies that are as large as can be practically shipped to the construction site. Shear walls and nonstructural partitions are often made of partially prefabricated assemblies. Steel columns are often delivered to the construction site in three-story-tall sections, and reinforcing cages and beam connections for hybrid structures are frequently fabricated on the ground before being erected (figure 27). HVAC equipment is premounted whenever possible on steel building girders before they are lifted into place. Japan's steel industry has recently developed a line of rolled steel shapes with constant flange depths, which helps speed detailing and aids in the standardization of the architectural and mechanical systems that must work around it.[19] The process of construction is often made more efficient through the design of buildings and construction methods that let workers perform exactly the same tasks each day, as if they were working in a factory. Mitsui's "One Day One Cycle" system is typical of such efforts (figure 28).

Construction Automation

The success of the car industry's automated assembly plants, combined with the construction industry's worker shortage, has helped spur the development of Japan's automated and robotic construction operations. Although the trend toward automation itself has produced some gains in productivity, the primary goal is to do a specific task with fewer people in a safer setting.

There are many construction robots in use in Japan,[20] including automated concrete finishers, steel-welding machines and facade inspection systems (figure 29). But these machines are not commonly used, and they often yield only incremental gains in both automation and safety. Although the comprehensives (which are leading the automation efforts of Japan's construction industry) have employed automated systems for about ten years, and are reported to have the "most advanced construction robots in the world,"[21] most of these systems automate one simple construction task and need a lot of supervision. The effectiveness of Japan's concrete floor-finishing robots, for example, is reportedly similar to American walk-behind systems and are "an order of magnitude" less effective than American riding trowels.[22] Neilson reports that generally the equipment "has not been successful" in increasing productivity.[23]

During a month-long trip to Japan sponsored by the Building Contractors Society (an industry group representing Japan's largest eighty or so Zenecons) in June 1991, only one working automated construction system was demonstrated at a job site: the Fujita Corporation's "Automatic Vertical Transport System" (AVTS), which was being used at their Makuhari Syokugyou Center Project (an office building). The system, which was still under development, was designed to deliver construction materials to work sites throughout the building. The system is installed after the building's superstructure is erected and works with materials delivered to the job site on standard-sized pallets. Once delivered to the site, human-controlled forklifts move the pallets to stacking zones next to an automated elevator system (figure 30). The elevator automatically loads the palletized materials onto the lift, delivers them to the floor where they are needed and stacks them on the floor. Once a set of pallets is delivered and stacked, human controlled forklifts deliver them to the location where they are needed. The system also works in reverse, removing unwanted materials from upper floors and delivering them to the ground floor or other floors where they are needed.

The elevator system is operated by one worker stationed at the ground floor, keeping him more comfortable and safe than he would be inside the lift. His only jobs are to tell the lift's computer control system which floor to take a given load of material to, and to direct the lift to pick up any materials which are ready to be sent to the ground floor. The operator's work is performed at a computer terminal, which displays an animated elevation-section of the lift's current location, the load inside it, and the status of the stacking area at each floor. A television screen next to the terminal displays the lift's interior and doors. If there is no room on a given floor to stack pallets, the computer terminal displays a blue box on the floor next to the elevator and will refuse to load any more pallets designated for that floor.

The most striking aspect of this system is its evolutionary, rather than revolutionary nature. Fujita claims that, when used to capacity, the system cuts manpower needs for material transport by more than half. They further note that since construction materials can be loaded on pallets at their point of manufacture, and not unloaded until they are about to be installed, the system minimizes the need for heavy lifting, producing further savings. The claimed savings are arguable for a few reasons: first, while the labor-savings produced by palletized transport are self-evident, the use of automated pallet transport systems is not new

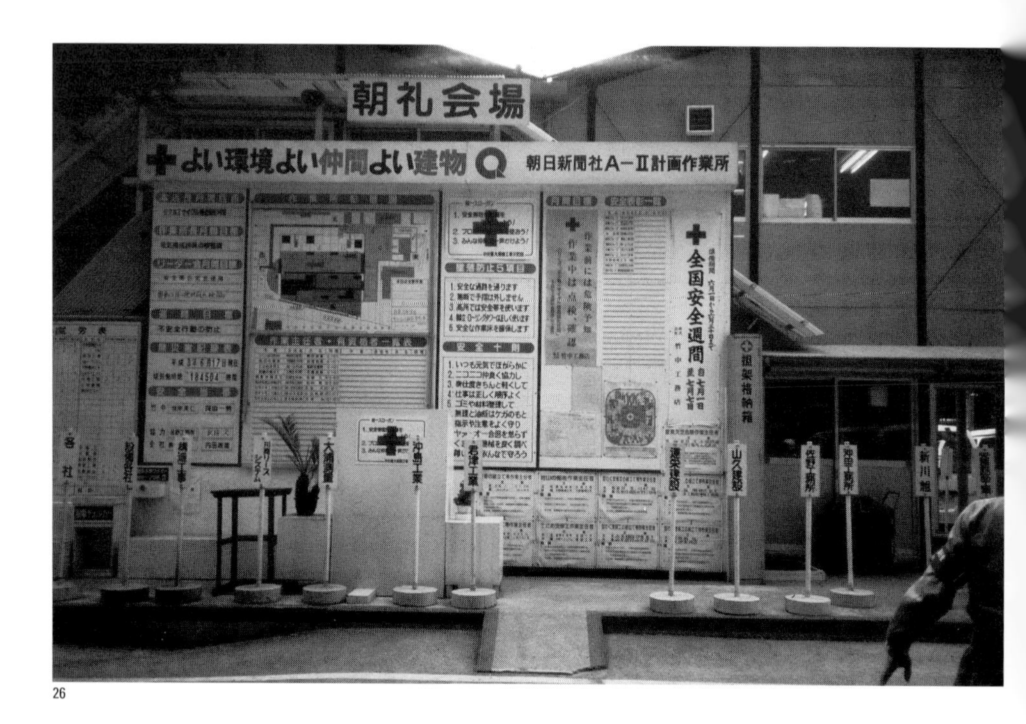

26

27

26. Safety posters at worker entrance to the Asahi Annex Plant Addition jobsite.

27. Reinforcing cage for cast in place column showing stub connections for steel girders.

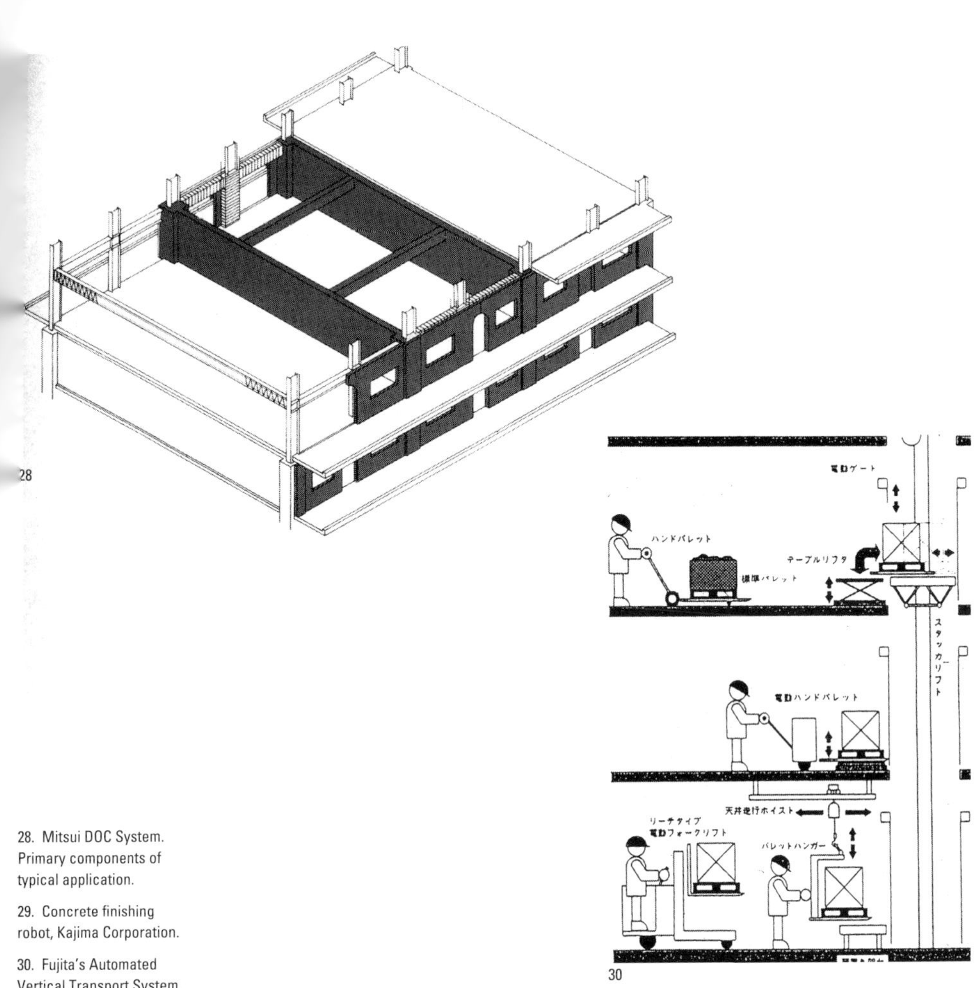

28. Mitsui DOC System. Primary components of typical application.

29. Concrete finishing robot, Kajima Corporation.

30. Fujita's Automated Vertical Transport System.

(a system developed for Ciba-Geigy is described below). Second, the materials delivered to and picked up from the lift need to be handled by human-operated forklifts. Finally, Fujita reports that only about 20 percent of the building's construction materials were handled by this system, due to its limited capacity (1 metric ton), and problems with delivery of materials on standard pallets. When all these issues are considered together it becomes clear that the system produces only small manpower savings.

Although the AVTS represents only an incremental improvement in automation, Fujita sees it as an essential component of an "Automatic Material Transport System" (AMTS) currently being designed. The AMTS envisions using unmanned forklifts that would receive goods from a staging area, and deliver them to the automated lift. The lift would take them to the proper floor, where they would be transported by unmanned forklifts to locations where they would be needed. Fujita claims that the AMTS is about two years away from testing at a job site.[24] Once operational, the system would be limited only by the capacities of its pallets, automated forklifts and elevators, and could be expected to produce significant manpower reductions and productivity increases.

Palletized, automated storage and distribution systems have already been used with great success in warehouse operations elsewhere. Ciba-Geigy, for example, has recently completed an automated, palletized warehouse in Milan.[25] In the Ciba-Geigy warehouse, every item stored fits onto a pallet, and an autonomous system of computerized machines controls all horizontal and vertical transport and tracks inventory. The warehouse was built around the pallet module, employing a regularly gridded structure resembling a huge library stack system. This regular grid greatly simplifies the problem of horizontal transport; Fujita's system, by contrast, would employ automated transport vehicles carrying material-laden pallets across the irregular and changing landscape of a construction site floor. Because of this, Fujita's completed system would represent a significant technological advance over today's warehouse systems.

Continuing Education

Because it illustrates the continuing education methods shared by all the comprehensives, the development process for the AMTS is as interesting as the system itself. The initial idea for the system came from Masahiko Sawaguchi, an engineer in Fujita's Transportation Department. Mr. Sawaguchi has been with Fujita since graduating from Muroran Institute of Technology in 1980 with a degree in architecture (and a major in structural engineering – see Chapter 4). Although he currently works at Fujita's office in Shibuya-ku, he learned much about the problems of construction during his first six years at the firm, which he spent at various construction sites. After gaining his job site experience, Mr. Sawaguchi moved to Fujita's design offices at Shibuya-ku, where he learned the basic principles of applied mechatronics and computer control systems in night classes taught by other Fujita employees. During this time he has occasionally been an instructor himself, sharing some of his specialized knowledge with other Fujita employees.

With this background Mr. Sawaguchi conceived a plausible, new automated system, even though he could not work out all the technical details himself. After obtaining approval to pursue the project, Mr. Sawaguchi oversaw its development, working with a design team comprised of Fujita employees and subcontractors (who supplied the computer control software and hardware). At the same time, he also continued to work on other projects. Mr. Sawaguchi reports that the development process took about one year, and during that time he was under much pressure from his supervisor to get the system working as fast as possible.

Mr. Sawaguchi's interdisciplinary training at Fujita, including his construction site experience and introduction to mechatronics, exemplifies a formal education process employed by all the comprehensives to extend and broaden the knowledge of their employees throughout their careers. Continuing education at Fujita, for example, can be traced through the career of a typical structural engineer. A typical employee in the Structural Division will almost always enter directly from a university, expecting to work at Fujita for life. He will spend much of his first two years working for extended periods at construction sites, and at either the Construction Management Division's home office or the R&D Division. Sometimes he will work at two or more nonstructural divisions during this time on domestic or foreign projects. For the next eight years, until his training period is considered complete, he will work under the close supervision of senior staff engineers. After about twenty years of employment, he will be expected to discern when and how to approach the R&D and Construction divisions for help solving special problems encountered in the projects he is by then managing. Of course, throughout his career, his education will be augmented by attending conferences held by other divisions and (if he chooses to review them) by keeping up with the in-house publications presenting their current work. Like Mr. Sawaguchi, this employee can also learn more through Fujita's in-house, nighttime courses.

This systematic investment in interdivisional communication and continuing education is used by the comprehensives to keep workers productive throughout their careers in an industry featuring constantly evolving technologies and methods. It is also the comprehensives' goal to foster technical and managerial innovations which might not emerge in a rigidly compartmentalized system. In the case of the AMTS, Fujita's investment seems to have paid off, in the sense that it promises new construction techniques which may significantly alleviate manpower shortages and improve job safety.

While Fujita's system seems to have benefitted both the company and Mr. Sawaguchi's career, the nature of the construction industry imposes limits to the benefits of automation. By comparison to the U.S., the amount of construction automation in Japan is significant, but its evolution has been slow compared to Japan's auto and electronics industries. This is mainly because the complex nature of building construction has hindered the rapid spread of robots seen in manufacturing settings. Manufacturing methods – based on the mass production of identical products – easily lend themselves to automation. In Japan's automobile factories, for example, cars travel along an assembly line as countless immobile robots build them piece by piece, each robot performing the same operation on each passing car. Automated machines typically perform more than 75 percent of the painting and welding work in these plants.[26] By contrast, all but the smallest buildings and their major components need to be constructed *in situ;* the scale and complexity of a building's structure, skin, mechanical systems and finishes make exact repetition in assembly operations rare. Equally important, each new building is unique in its *parti* and dimensions. So, a construction firm wishing to automate its operations needs to build mobile robots which perform a set of similar, but not identical tasks. This is complicated by the fact that reprogramming the equipment is required for each new building, and sometimes each new floor. These requirements make automation much more difficult to achieve, which is reflected in the different amounts of automation found in Japan's construction and auto industries.

Building Use and Maintenance

The comprehensives use a wide range of technologies to improve a building's "amenities" or "software." Amenities or software refer generally to anything that improves the quality of life in a given space. To help market particular amenities, the comprehensives' sales staffs often link them to increasing happiness and workplace productivity, which makes the software more attractive to corporate clients.[27]

Most of the comprehensives' efforts to improve a building's user-friendliness fall under the nebulous category of "intelligent building systems." Although there is generally no consensus among designers about what an "intelligent building" is, it has been usefully defined as a building that " 'knows' what is happening inside and 'decides' the most efficient way to provide a convenient, comfortable, and productive environment for its users."[28] Using this definition, intelligent building systems have been used in Japan and the West for some time. The Intes Building's AMD and photosensitive blinds are examples, as are the noise control and tuned mass damper systems employed at the Citibank Building in New York. But while most Western intelligent building technologies have been limited to improving structural, HVAC, lighting and acoustic control systems, the Japanese also employ intelligent systems to replace many manually provided services to improve a building's "amenity level."

The Makuhari Techno-Garden Complex, a set of two buildings located in the Makuhari New Business District in the Chiba Prefecture outside Tokyo, shows the comprehensives' use of intelligent building systems to improve the building's environment. The complex, which includes two twenty-four-story towers connected by a six-floor office block, was designed and constructed by the Shimizu Corporation for Makuhari Techno-Garden, LTD., and was completed in March 1990 (figure 31). The buildings are leased primarily by high-tech industries that are developing new technologies. To minimize the amount of space each company has to rent in this expensive complex, Shimizu designed a set of common meeting rooms that are shared by all the tenants. Tenants wishing to schedule a meeting in a common room reserve time and space on an owner-supplied personal computer (PC), connected via a local area network (LAN) to the landlord's central mainframe. The desired room and meeting time are logged into the computer, which automatically arranges (with human security personnel) to unlock the space at the designated time, sets the central heating, cooling and air-conditioning system (HVAC) to moderate the room's climate during the meeting, and automatically invoices the tenant. A mainframe computer is used to control the complex's entire security, environmental control, billing and telecommunications services. Telecommunications service includes private branch exchange (PBX) and voice-mail systems for each tenant, wide-band data transfer (used for high-speed computer-to-computer communications), and picturephone capability. The building's environment is moderated by computerized HVAC systems that control local environments as a function of the particular machines and people occupying them. Sensors monitor occupancy and turn lights on and off automatically as one moves through the building. Shimizu's environmental control systems extend beyond temperature, humidity, noise and light levels, to include smells. At the tenant's option, Shimizu will install its proprietary olfactory control system, which will exude various scents (including, for example, nutmeg, basil, and peppermint) at preprogrammed times (figure 32). Shimizu claims that the exhausted smells effect worker happiness and productivity. The company also brings this total-control approach to Makuhari's waste management problems. The

31

32

33

31. Makuhari Techno-Garden, Shimizu Corporation. General View.

32. Olfactory Control System. Shimizu Corporation.

33. Fujita Corporation Headquarters, Tokyo. Automated facade- cleaning machine.

complex features vacuum-controlled garbage disposal and grey water recycling systems.

The occupancy-monitoring sensors are used for security as well as environmental control. Whenever someone enters an unoccupied space, the sensors inform the mainframe, which checks to see if the tenant has authorized occupancy at the time. Human security personnel are notified if there is a possible problem. To improve security and make life more pleasant for tenants, the entire Makuhari complex can operate on a cashless basis. Employees have "smart" ID cards that they can use to purchase anything sold in the complex's first-floor shops. To encourage their use, items purchased with a card are discounted 7 percent. These cards are also used throughout the complex instead of keys, to control access by space and time. A computer programmer working for an electronics firm, for example, will have access to his or her workroom at all times, while a janitor will be allowed access to the same area only at night. The janitor's card will open the supply closet while the programmer's will not, and neither employee's card will open the door to the firm's "clean" manufacturing room, which is accessible only to those directly responsible for working with the sensitive equipment inside.

The comprehensives also look to technology to help make their buildings easier to maintain. Maintainability is usually improved through direct automation of manually performed tasks, rather than by a network of interrelated, intelligent systems. This approach echoes the comprehensives' strategy for construction automation. The upkeep systems employed by Fujita and Kajima illustrate some of Japan's automated building-maintenance techniques. At Fujita's headquarters building in Tokyo, an unobtrusive metal track is laid between each vertical joint in the building's tiled facade panels. An automated facade-washing machine rides up and down along the track (figure33). Fujita uses the system to clean almost all of the building, which Mitsubishi Heavy Industries developed in the 1970s. Robots arrange furniture in several of the building's meeting rooms and store it when the rooms are being cleaned or used for functions requiring open space. Fujita also markets robotic floor-cleaning systems for office interiors. Finally, Kajima uses a tile-inspecting robot to locate areas of tile-faced facades needing repair. This robot saves money on facade inspection, which is otherwise done manually by engineers on scaffolds.

Although the Makuhari complex and Fujita's Headquarters use more automated maintenance and operation technologies than are found in comparable Western buildings, their intelligent systems are mostly based on proven technologies, previously used elsewhere.[29] What is fascinating about these buildings, especially Makuhari, is not their individual intelligent systems but the fact that their designers and builders have marketed them as a comprehensive package for simplified building use and maintenance. By incorporating convenience, security and maintenance systems in their structures, the comprehensives go well beyond a builder's traditional role of providing shelter, and move into tenant service areas that are usually addressed by an *ad hoc* constellation of service companies and consultants. As they have expanded the domain of designer-constructor, the comprehensives have deployed an unusually large number of operational technologies in individual buildings. During the 1980s, when new building clients sought the prestige of working in more futuristic spaces than their rivals, this led to expanded expectations of the services a building *should* perform, and increased demand for the comprehensives' maintenance and amenity products.[30] In Japan's more frugal "post-bubble" economy, continued corporate interest in more high-tech gadgetry has yet to be evinced, and this particular "demand-creation" strategy may not continue to work as well.

Table I: Typical Vibration Control Technologies Installed in Japan

Building	Engineering/ Construction	Location/ Year Completed	Technology
Sendagaya Intes	Takenaka Takenaka	Tokyo 1991	Active Mass Damper
Kyobashi Seiwa Building	Kajima Kajima	Tokyo 1989	Active Mass Damper
Nagasaki Airport Control Tower	Shimizu Shimizu	Nagasaki 1988	Tuned Sloshing Damper
Crystal Tower	Takenaka Takenaka	Osaka 1990	Tuned Mass Damper
Seavans	Shimizu Shimizu	Tokyo 1991	Elastic-Plastic shear wall
Taisei Technology Research Center	Taisei Taisei	Yokohama 1988	Base Isolation
Edogawa-ku	Kumagai-Gumi Kumagai-Gumi	Tokyo 1988	Base Isolation
Toushin 24 Omori Building	Kajima Kajima	Tokyo 1990	Base Isolation

Source: Building Contractors Society, Tokyo.

Chapter 2 Endnotes

1. Author's interview with Mr. Jin-Ichi Yoshida, Taisei Corporation, June, 1991.

2. The comprehensives also build designs that have been completely prescribed by independent architects; their commissions in these cases are similar to a contractor's in the American bid system.

3. While the strength of a standard component is measured by the load that causes it to break, its ductility is measured by how much energy it absorbs.

4. Author's interview with Yoshida, June 1991.

5. R. Tucker, et al., ed. *JTEC Panel Report.*

6. All mechanical vibration control systems use some non-load-bearing device to dampen a building's oscillations caused by wind or earthquakes. Active dampers employ motion sensors to measure a building's wind or earthquake-induced movement. This information is fed instantaneously to a computer, which directs a set of hydraulic jacks or other equipment to deform the structure in the opposite direction.

Rather than actively adding new forces to counter a structure's movement, passive dampers reduce vibrations by shifting some of the structure's kinetic energy into heat or the vibrations of a nonstructural mass. Tuned mass, viscoelastic and other passive damping systems, although not as efficient as active dampers, are less expensive, need less maintenance and require no external power source.

Tuned mass dampers (TMDs) work by fastening a mass-block to a structural component (such as a column) via a spring and a damping device (which is similar to a shock absorber). This system is set up so that the building's own resonate-frequency vibrations (which could be caused by wind or earthquakes) induce analogous movement of the mass block and spring. By the conservation of energy, the TMD motion in turn reduces the amplitude of the building's vibration. The damping device also increases the TMD's effectiveness over a range of frequencies and takes a small amount of mechanical energy out of the system as heat. Since each TMD is "tuned" to a particular resonant frequency, individual TMDs need to be installed for each resonant frequency at which a building may vibrate.

For an in-depth analysis of TMD systems, see: A. Webster and R. Vaicaitis, "Application of Tuned Mass Dampers to Control Vibrations of Composite Floor Systems," *Engineering Journal*, third quarter (1992): 116–124.

7. T. Tamura, K. Fujii, K. Sato et al., "Wind Induced Vibration of Tall Towers and Practical Applications of Tuned Sloshing Dampers," *Proceedings, Serviceability of Buildings Workshop* (Ottawa: University of Ottawa Press, 1988).

8. For an in-depth description of tuned mass and active dampers, see:

A. Webster and M. Levy, "A Case of the Shakes," *Civil Engineering Magazine* (February 1992).
9. M. Schmertz, "Citicorp Center," *Architectural Record* (June 1978).
10. *JTEC Panel Report*, 44.
11. Author's interview with Sato Chikafusa, June 1991.
12. K. Reinschmidt, F. Griffis and P. Bronner, "Integration of Engineering, Design, and Construction" *Journal of Construction Engineering and Management*, vol. 117, no. 4 (December 1991).
13. As discussed above, these documents were almost exclusively created from the building's base structural and architectural documents, which were of course completed before construction began.
14. R. Lubben, *Just in Time Manufacturing: An Aggressive Manufacturing Strategy* (New York: McGraw Hill, 1988), 44–46.
15. T. Kidder, *The Soul of a New Machine* (Boston: Little, Brown & Co., 1981).
16. Civil Engineering Research Foundation, *Transferring Research Into Practice*, chap. 3.
17. "The Fire This Time," *The Economist* (October 10, 1992): 36.
18. *Engineering News Record* (March 15, 1990): 23.
19. J. Fisher, *JTEC Panel Report*, 103.
20. S. Levy, *Japanese Construction*, 283.
21. *International Symposium on Construction Automation*, Findings reported by Obayashi Research & Development (1990).
22. R. Neilson, *JTEC Panel Report*,64.
23. R. Neilson, *JTEC Panel Report*, 64.
24. While this may be an optimistic prediction, it now seems more plausible given the existence of Fujita's automated furniture setup system, a fully operational system for arranging furniture in large, multi-use spaces, featuring robotic forklifts similar to those needed for the comprehensive construction transport system.
25. "Centro Distribuzione Merci: Il Futuro Nei Robot, " *Abitare*, vol. 227 (September 1984), 36.
26. Author's interview with Nissan Motor Corporation representative, July 1992.
27. Fujita Corporation, *For the People – In Search of Amenities*, Report on new headquarters building (June 1991).
28. J. Bennett, et al., *Capitol and Counties Report*, 62.
29. "Making Buildings Smarter," *Engineering News Record* (November 1, 1984); Ken Senior, "Emerging Technology," *Building Design and Construction* (November 1989), 76–80.
30. "That Certain Japanese Lightness," *The Economist* (August 22, 1992), 75–76.
31. Author's correspondence with the Building Contractors Society, March, 1994.

Chapter 3

Technology Tomorrow: Research and Development

The Scope and Role of the Research and Development Divisions

The most striking feature of the comprehensive construction companies is their large, well-staffed and funded research and development (R&D) divisions. Unlike their American counterparts, which generally do not have sizable R&D operations, the comprehensives are committed to R&D as both a means of increasing their short-term profitability, and as an essential part of their long-term growth strategy. The R&D operations of the Big Six firms are particularly impressive.[1] Their research facilities are typically "larger and better equipped than the largest university-based labs in the U.S. or elsewhere"[2] (figure 1); their funding averages almost 1 percent of sales, compared to America's 0.04 percent[3] for firms of similar size; their staff is mostly drawn from graduates holding Ph.D. or masters degrees from Japan's best universities; and their work is generally of a high quality, as measured by, for example, the frequency with which their research is published in prestigious American engineering journals. R&D divisions are not only flourishing in Japan's Big Six firms – virtually all of Japan's large general contractors have R&D divisions with their own permanent facilities, budget and full-time staff. The R&D capabilities of Japan's twenty-seven largest comprehensives are summarized in Table I.

To a Westerner accustomed to new design methods emerging from universities and consulting engineering firms, and the development of new building products by manufacturers, the presence of R&D facilities in Japan's large construction firms is surprising. That the comprehensives are pursuing building design and construction R&D in particular is especially perplexing. Unlike Japan's civil engineering construction industry (in which, as described in Chapter 1, the comprehensives also feature prominently), its building construction business is dominated by negotiated contracts that emphasize long-term client relationships over competitive bidding. A client's decision to engage a particular firm to design and construct a large-scale building is often based on the firm's previous work for the client and, in the case of large corporate clients, there is pressure on them to choose a contractor from within their *Keiretsu* umbrella. (A *Keiretsu* is a group of conglomerates, led by a bank, that controls aspects of an industry by functioning as an oligopoly). This system could be expected to foster corporate complacency among the

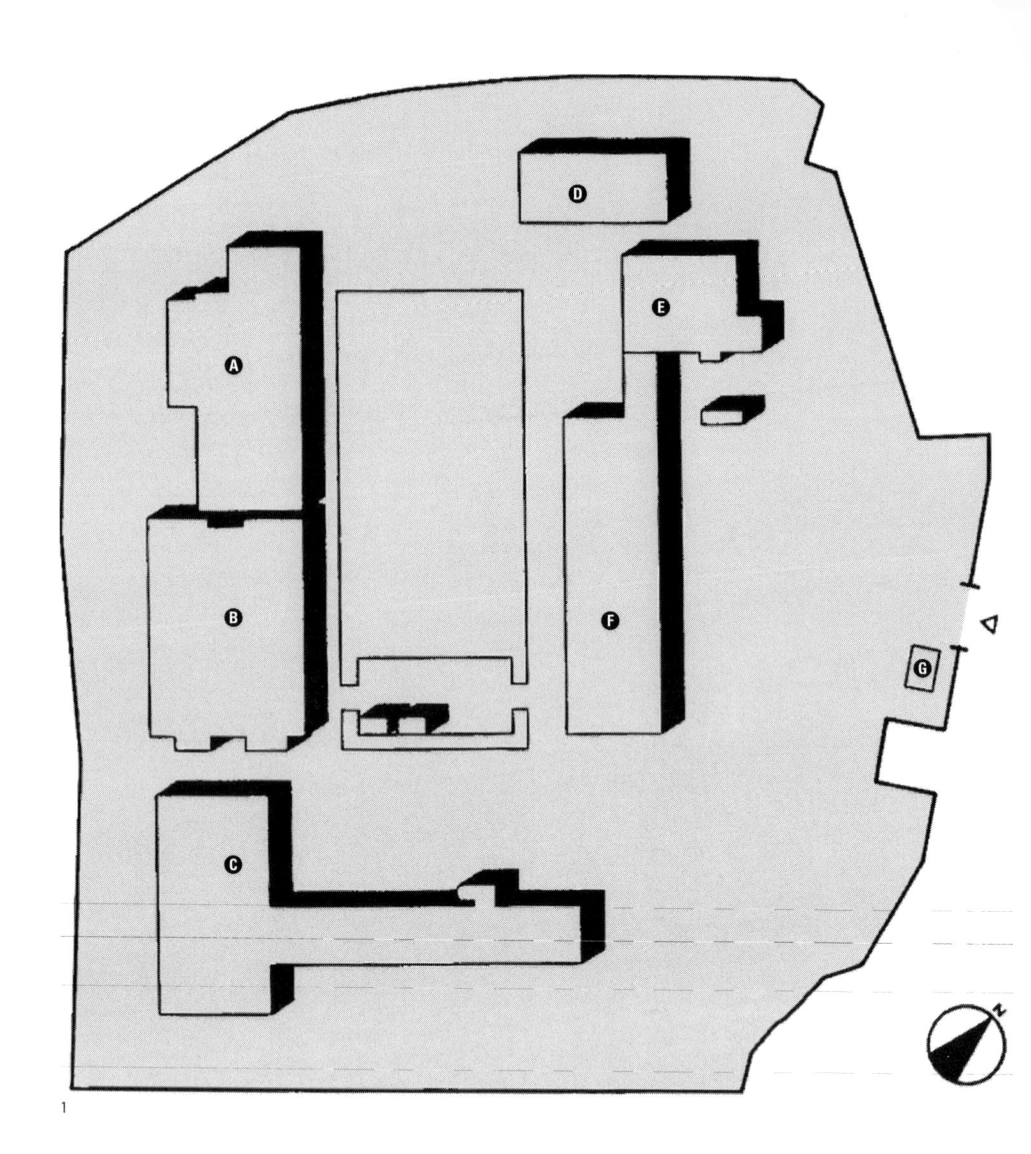

1. Taisei Corporation, Technology Research Center, Plan.
A. Hydraulics laboratory
B. Materials laboratory
C. Administration building
D. Environment laboratory
E. Acoustics laboratory
F. Structural laboratory
G. Guard house

contractors instead of the competitiveness that drives the comprehensives' building R&D efforts.

The reasons for the comprehensives' commitment to R&D are many and complex, but most important is their shared belief that obtaining new technologies, over the long haul, is as important to maintaining a competitive edge and continued growth as nurturing existing clients. The essential role of R&D in the comprehensives' long-term corporate strategy cannot be overstated. Sadayasu Asano of the Shimizu Corporation reflects a typical corporate view when he says that "only those companies capable of offering new technologies and services through R&D will continue to exist."[4] R&D's importance to the comprehensives underscores the large differences between Japanese and American notions of what a construction company should be. Unlike American firms that tend to view themselves as implementors and installers of technologies developed by others, Japanese contractors take pride in developing their own technologies for improved building materials and aiding the construction process.

The comprehensives' commitment to technological development has its roots in Japan's reconstruction period following World War II, when the first of their R&D divisions were founded. While rebuilding Japan after a war that destroyed 40 percent of its urban structures and left it with very little capital, the construction firms were under great pressure to build quickly and efficiently. During the war, successful Western corporations had shown how R&D could help reinvigorate an industry. American fighter planes, for example, were arguably the worst among industrialized nations at the onset of the war, but largely due to successful R&D efforts, were by 1945 almost unsurpassed. The benefits of corporate R&D were quickly recognized by Hazama-Gumi, which initiated an R&D division in 1945.[5] The devastation of the war also afforded Japan's industries, including its construction firms, an opportunity to modernize their corporate structures as they rebuilt their facilities. American industry's use of "technological development [to] enhance the competitive power of major companies,"[6] and the cardinal role that R&D divisions played in their long-term growth strategies, was not lost on Japan's auto makers or construction companies. Besides Hazama-Gumi, some of Japan's largest construction firms, including Obayashi, Kajima and Shimizu, were quick to embrace this ideology and organized R&D divisions in the late 1940s. To remain competitive and retain their longtime clients, most of the other large firms felt that they needed to establish similar facilities, and also initiated R&D programs in the following decade.

The Japanese penchant for high-tech, high quality products is also an important influence on the construction industry, and is one of the reasons why R&D is deemed so important. Japan's retail consumers have an almost insatiable appetite for the latest technological equipment. This love affair with technology manifests itself in many ways, from the huge outpouring of high-tech camera and other electronic products in Tokyo, to the tradition of trading in new cars that often have been driven less than 12,000 miles.[7] Corporate customers thinking of constructing a new building have a similar affinity for technology and want their new structures to be as up-to-date and technologically advanced as they can afford. Intelligent building systems and other technologies are a prominent feature of the comprehensives' marketing efforts. In this context, new building systems developed by their R&D staffs help the comprehensives hold on to their long-term clients and attract new ones.

As R&D is used to help tout the high-tech image of the comprehensives, it is also used to help counter the image of the construction industry as "dirty, dark, and dangerous" – the so-called 3 K's. In the minds of many prospective engineers and construction workers, the construction industry lacks the glamour of traditional high-tech industries, and its manual labor compares poorly to other entry level, service-sector jobs available to an increasingly well-educated work force. R&D projects are used by the comprehensives both to add cachet to the perception of the industry among university students and to reduce the need for dangerous, dirty and strenuous work at their job sites. Alluring R&D projects for space exploration, kilometer-high megastructures, and deep-underground cities are used by many comprehensives to help attract graduates from Japan's universities. Various automated construction methods (see Chapter 2) are being developed to reduce the number of difficult and unpleasant manual construction tasks. Robots also make construction seem like a more high-tech, and therefore more appealing field.

Although the federal government plays a role in encouraging and directing the R&D efforts of Japan's construction firms, it does not wield power over the Zenecons comparable to that of the Ministry of International Trade and Industry (MITI) over the auto and electronics industries. The Ministry of Construction (or MOC – described in detail in Chapter 4) is the federal body largely responsible for overseeing Japan's construction industry. Like its power over the industry it regulates, MOC's industrial strategy also differs from MITI's. While MITI has nurtured the development of a few select firms, MOC has encouraged the spread of technologies to many construction firms. This strategy fosters competition among many of the largest firms, and encourages the development of new products by smaller companies as a way to improve their position in their industry.

While these forces have encouraged the development of formidable R&D divisions among the comprehensives, they have also helped create a sameness in their operations. To ensure that they do not fall too far behind in any important potential growth area, most of the comprehensives try to pursue R&D projects in the same fields as their competitors. Today, virtually every comprehensive construction company is pursuing a number of core areas of research, such as construction automation, high-strength concrete, earthquake studies, and intelligent building systems (Table II). Jin-Ichi Yoshida of Taisei Corporation, says that the similarity of R&D efforts among the comprehensives is also due, at least in part, to "the lack of individualism in Japan."[8]

The similarity of the comprehensives and particularly the similarity of their R&D efforts, stands in marked contrast to the situation in the U.S., where construction firms have not generally been perturbed about their competitors' moves to fill a particular niche, and have often used a strategy of niche-filling to stay competitive. Bechtel, one of the few American companies with a large research staff (including approximately 300 employees), has recently focused much R&D work on hazardous waste disposal technologies, considering it a growing market through the nineties. M. W. Kellogg, one of Bechtel's American competitors that also has an R&D group, has chosen instead to focus on new technologies for building and maintaining large chemical plants.

Besides working in the same areas, the comprehensives' R&D divisions share conceptual goals. The purpose of much of the comprehensives' R&D efforts (as they relate to building design, construction, use and maintenance) is to decrease the time and cost required for design and construction, to increase the quality of a building and improve the environment for its occupants. These priorities are pursued in both the project support and product development functions of the R&D divisions.

Project Support

Facilitating the completion of extant design and construction commissions is one of the comprehensives' R&D divisions' more important functions. Although this "project support" accounts for only about 20 percent of a typical R&D division's work,[9] its impact on the firm's main business is substantial. The comprehensives call on their R&D staffs to both improve the design and expedite the construction of their major commissions. As outlined in Chapter 2, R&D divisions may plan and execute full-scale testing of various portions of active building commissions. Depending on the nature of the project, the role of an R&D section may also take many other forms. How this works is probably best understood by looking at a few examples.

The design process used by Takenaka in the 552-seat Hamarikyu Asahi concert hall of the Asahi Newspaper Annex addition (described in detail in Chapter 2) exemplifies the comprehensives' use of in-house R&D technology to improve a project's design. After deciding on the size, approximate footprint and volume of the hall space, the project architects turned to the R&D division to provide a specific interior shape and a set of appropriate materials to maximize the acoustic performance of the hall. Drawing on the basic hall parameters and their past experience, the R&D staff began by preparing a trial acoustic design. Using a proprietary acoustic-modeling computer program called STRADIA, developed by Takenaka, researchers compared the performance of the trial design with the previously recorded performance of famous concert halls in Vienna, Salzburg, and Zürich (figure 2). Differences between the analytically predicted sound of the Asahi hall and the existing spaces were used to improve the shape and performance of the trial design. In an anechoic chamber, the resulting optimized, computer-generated design was verified by human listeners. This was done by arranging a series of speakers in a circular pattern, and by controlling the sound from each with the STRADIA program. The STRADIA system made the sound inside the chamber approximate various parts of the as yet unbuilt Asahi hall (figure 3). When the design was complete, the client was called in to listen to the computer-simulated performance of the hall in the anechoic chamber, and to approve its acoustic properties. As a final test of the STRADIA program's accuracy and capabilities, Takenaka plans to record the sound quality in the completed Asahi hall and compare it with the computer simulated design.

In the engineering design and construction of Norman Foster's Century Tower in Tokyo, Obayashi's structural and mechanical engineers called on the R&D division to help work out some complex design problems, and to help obtain government approval of a previously untried design. Foster's architectural design for the typical office spaces of Century Tower employs column-free floors alternating with mezzanines hung from them, both opening onto a central atrium (figure 4). Sectionally, the spatial concept of the building relies on a very open relationship between the working floors and the atrium space. This design left no room for the type of fire-shutter system that Kajima employed in its atriumed headquarters building in Tokyo (see Chapter 2). The conceptual solution to Foster's problem was proposed by Obayashi's Research and Design Divisions. Their schematic design featured fire separations using high-pressure water jets at the edge of every other floor (directed at the floor two stories below), that would turn on during a fire. These "water curtains" would effectively separate each working floor from the atrium and each two floors from those above and below (figure 5). In this scheme, smoke would be contained by barriers suspended from the ceiling that would automatically rotate downward into position when the local fire alarm system was triggered. Smoke would be exhausted via special returns (separate from the normal air-handling system); the atrium would be maintained at a higher pressure

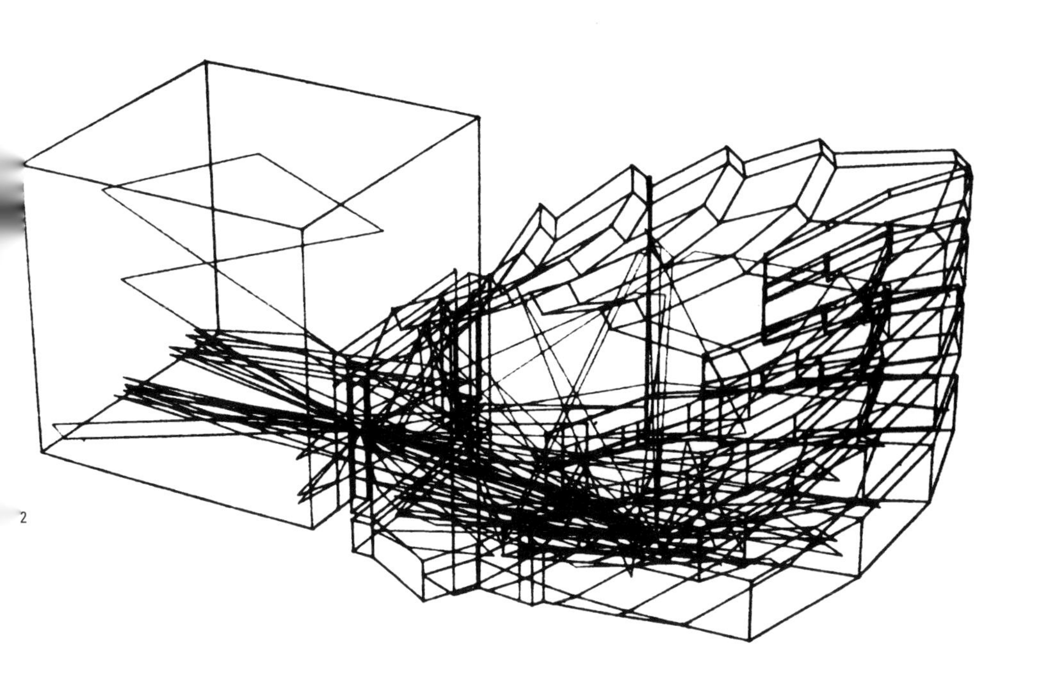

2. Computer-generated reflected ray tracing diagram.

3. Computer-simulated sound in anechoic chamber.

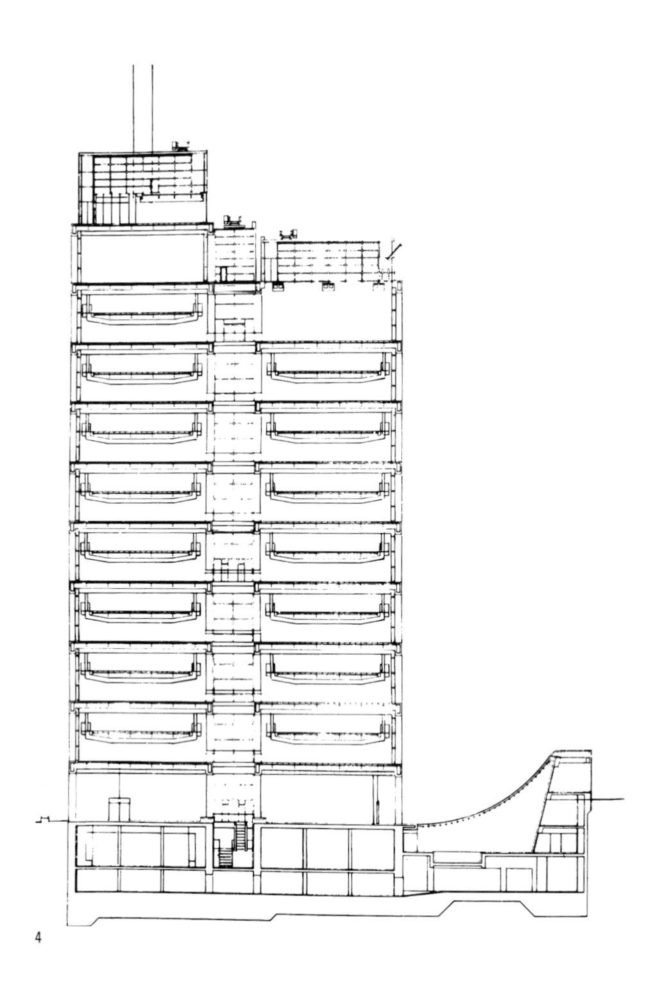

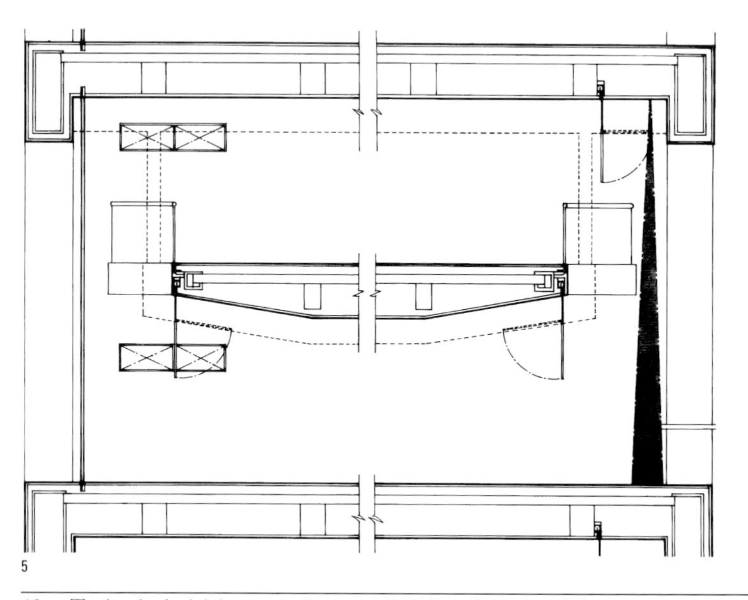

4. Century Tower. Building Section.

5. Century Tower. Fire containment system. Section.

than the adjacent occupied space by directing the atrium's HVAC system to close its air exhaust ducts while continuing to provide supply air.

J. R. Preston, the mechanical engineering firm retained by Foster to schematically design the building's environmental control system, worked with Obayashi to develop the system. Working with their R&D division, Obayashi's designers completed the system's design, prepared construction documents for it, and submitted it for government approval. During the approval process of this untried system, the government dictated that a scale model be prepared and tested. The company's R&D staff was again called on to prepare and test a one-tenth scale mockup of three floors of the building.

The R&D division also facilitated the analysis and government approval of the building's structural design. To meet its earthquake requirements, the Japan building code requires a complex, elastic-plastic time history analysis be performed for any building of more than 60 meters. (The phrase "time history analysis" means in this context that a computer model of the building's structural system is analytically subjected to an earthquake, calculating the structure's predicted response every fraction of a second. The "elastic-plastic" clause indicates that the analysis must properly account for the behavior of portions of the building that have yielded and can essentially take no more load.) Having previously designed some high rises in Japan, Obayashi of course had performed these analyses in the past. But the plan of Foster's building and the different heights of its towers provided particular problems not encountered previously by Obayashi.[10] The nonsymmetric geometry of the building was such that, when subject to earthquake motions along a north-south line, for example, the building would not just sway north and south but also twist. This complicated the analysis of the building enormously, which could otherwise have been approximated with a two-dimensional model. In this case, members of Obayashi's R&D staff developed the theoretical underpinnings of a new computer code for modeling the building's three-dimensional behavior.

The interdivision team approach used to develop the 3-D computer code for Century Tower is typical among the comprehensives. While the theoretical aspects of the code were developed by members of the R&D staff, Obayashi's computer support division wrote the algorithms and the computer code itself. Once the code was written, the dynamic analysis of the structure was performed by the Century Tower design staff.

The comprehensives also use their R&D divisions to help with construction problems and to test the validity of new construction methods for particular jobs. For example, in the construction of a 100-meter-tall reinforced concrete building in the Shikoku Island region, Toda Corporation needed to construct a series of complicated, highly reinforced moment connections using high-strength concrete. Because Toda had not previously used concrete with this strength (4.2 KN/cm^2) for such a complex, large connection, the construction division asked the R&D staff to prepare a full-scale prototype connection. Working with the construction staff for the building, the R&D division developed the specifics of the reinforcing pattern to ensure that it could actually be built and ensure that the concrete at the joint could be poured and vibrated in a way that avoided the formation of voids (air pockets). After the prototype was constructed, core samples were taken to verify the quality of the placement techniques, and to compare the concrete strength with that predicted for the high-strength mix.

While none of these examples reveal any revolutionary analysis techniques or design/construction methods, they do show that having R&D support in-house improves the comprehensives' design and construction efficiency and helps them achieve a high level of quality. The type of acoustical modeling performed by the STRADIA program, for example, is well known. In the U.S., however, this technology resides mostly at universities or with specialized acoustic consultants. University professors, whose main concern is basic research, often find it difficult to focus on building design problems, and to work efficiently within a construction project's tight schedules. Specialized consultants can do this work, but they too have their own priorities and schedules, making it hard for them to pay full attention to a particular project if other work is pressing. Their commitment to a project's quality is also arguably not as deep as those responsible for the overall building design and construction. By having all the support staff it needs under one roof, a comprehensive ensures a timely response to exigent design and construction issues, and easily controls the quality of its project support work.

An obvious disadvantage of this system is the need to maintain both project support staff and facilities, even when they are not needed to work on a specific project. In addition, as new equipment becomes available (such as more powerful computers), it is harder to justify obtaining it while in-house machines are still being depreciated. Similarly, if the capabilities of its staff become less sharp over time, the comprehensives cannot replace them as easily as they could a consulting engineer or university professor hired on a project-by-project basis. Some comprehensives are working to solve these problems by encouraging their R&D staffs to provide services like wind tunnel testing and computer modeling to independent architects and engineers. This strategy can help the R&D divisions through slack times at their own firms, and encourage them to stay technically current in an effort to attract new, noncaptive customers.

Product Development

The comprehensives' R&D groups justify themselves primarily through their product development work, which takes most of their time. This work is interesting not only because it is typically performed at universities or by manufacturers in the West, but, more importantly, because the broad scope of these efforts is considered to fall outside the purview of the mainstream American construction industry. The comprehensives do not restrict themselves to the development of new structural methods, building materials and components, although these account for much of their R&D initiatives. Beyond studying new "hardware" (structural, envelope and HVAC systems), the comprehensives' R&D divisions are also developing ways of improving a building's amenities or software. As noted in Chapter 2, amenities or software are building systems that address the quality of life and productivity of a building's occupants. R&D work in building amenities by the comprehensives includes, among other things, relationships between environment and human comfort, clean room technology, personal microclimate workspace controls, olfactory systems, personal information management and communications systems (Table II). The comprehensives' work in these areas is consistent with their corporate marketing ideology: to provide customers with a complete product, including habitable space and all the systems required to make it fit the client's particular needs. This philosophy leads them to compete in many software arenas, and, as described in Chapter 2, encourages them to develop new software technologies to enhance building use and maintenance.

The interior atmospheric research being done by Taisei's R&D division is a good example of how these firms are working to improve building amenities.[11] Research on occupant comfort levels, including the effects of humidity, temperature and local air velocity, is being done at Taisei's Technical Research Institute. Charts describing zones of comfort, based on temperature and humidity, have long been available throughout the world for use in designing environmental control systems, but these are largely based on work done in the U.S., and Taisei's research has shown that Japan's climate and traditions give its people different notions of comfort. Additionally, research done in Denmark, and subsequently verified by Taisei in their office mockup labs, shows that comfortable temperature and humidity levels can be extended by controlling air velocity.

This basic research, which would most probably be conducted at universities in the U.S. (and would certainly not be performed by an American construction company), is being used by Taisei as a basis for the development of new environmental control systems. According to Taisei, besides "actualizing comfort for our customers,"[12] the new systems will include air-conditioning equipment (designed as a joint-venture with an air-conditioning equipment manufacturer) that will reduce the energy required for cooling by increasing the air velocity in office spaces.

As described in Chapter 2, AMDs have recently been installed in buildings by the comprehensives to control wind and earthquake induced vibrations. Obayashi's research on active dampers is typical of the work being performed by the comprehensives to improve this technology. The company is developing a system it calls a dynamic vibration absorber (DVA), to function as "an active vibration control system which curbs vibrations in medium and small scale earthquakes and under strong winds ... [and] improves residential conditions of high-rise buildings."[13] Currently the system's sensors, servo-motor and computer controls are being tested on a full-scale, two-story vibration-research building at Obayashi's Technical Research Institute in Tokyo (figures 6 and 7). Preliminary results suggest that the system can optimally reduce building displacements due to earthquakes by 75 percent. The R&D staff is currently working with the firm's building design engineers on a design for a commercial structure, which Obayashi plans to install within the next few years.

Apart from AMDs, the comprehensives are working on several experimental components and construction techniques aimed at reducing structural costs. Shimizu, for example, has recently developed a steel-encased, radially constrained, unbonded, reinforced concrete column system it calls a "super concrete column" (figure 8). By radially restraining the column's concrete, the steel tube encasement produces a column with much more strength than a standard reinforced concrete column of the same diameter. These columns have recently been used by Shimizu in their Seavans complex in Tokyo to carry heavy vertical loads with smaller diameter columns than traditional reinforced concrete technology would allow.

Taking a corollary tack, Toda's R&D division is developing unbonded, concrete-jacketed steel struts (figure 9) which use a precast concrete sleeve to keep the strut's steel from buckling. In this way, the amount of relatively expensive steel required to carry a given load is reduced. Because the concrete provides fire protection, the assembly also benefits from a reduction in the required amount of site-applied fireproofing.

To cut the cost of providing lateral earthquake resistance in reinforced concrete structures, Obayashi corporation has recently developed a hybrid precast-panel, poured-in-place frame shearwall system (figure 10). Besides saving on forming and site-pouring costs by using the precast infill panels, the system is further simplified by rebar connections between only the panels and the beams above and below them, not also between panels and adjacent columns, as is usual. By testing large-scale, three-story mockups of the system, Obayashi found that top-and-bottom-

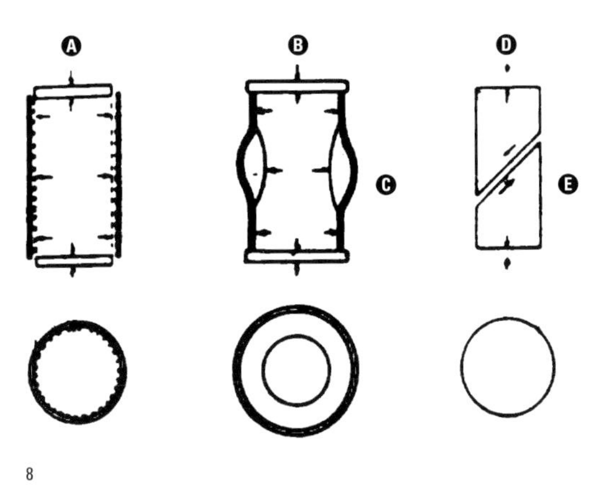

8

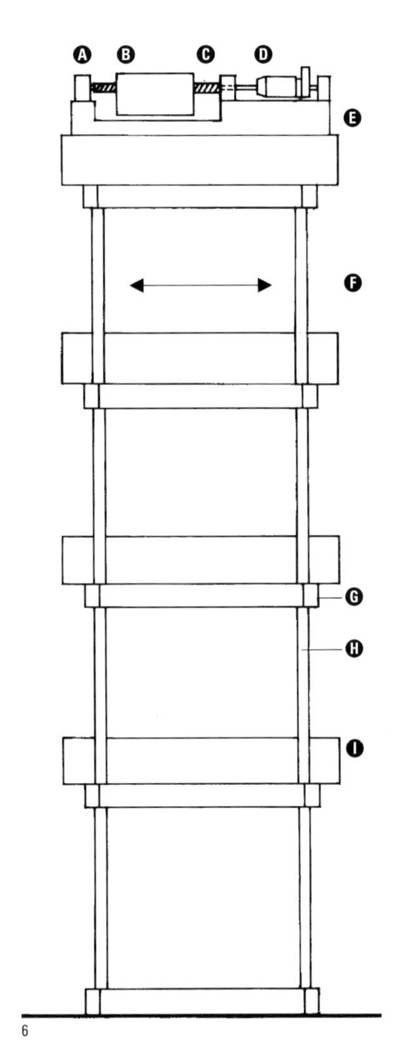

6

6. Obayashi Dynamic Vibration Absorber (DVA).
A. Bearing
B. Mass
C. Ball screw
D. AC servo motor
E. Active mass damper
F. Direction of vibration
G. Steel beam
H. Steel column
I. 2.5 ton mass block

7. Obayashi DVA installed on building.

8. Shimizu's Steel Jacketed, Radially Constrained Reinforced Concrete Columns.
A. "Super Concrete Column" test
B. Axial force
C. Localized steel yielding and concrete crushing
D. Unconstrained column
E. Shear failure

7

9

10

9. Testing of Toda's unbonded, concrete-jacketed steel strut.

10. Elevation. Obayashi precast-panel / poured-in-place frame shearwall system.

only panel connections helped increase the ductility of the system far beyond what would normally be expected for shearwalls.[14] After refining the system through further tests and computer analyses, Obayashi achieved similar strengths as those found in continuously reinforced shearwalls, without losing the ductility of the system.

Although each Japanese firm has its own particular process, the evolution of Obayashi's hybrid shearwall system illustrates how the comprehensives' R&D product development efforts unfold. Because of their low ductility, reinforced concrete shearwall systems are generally frowned upon by the government for high-rise buildings. In 1982, thinking that it could produce a system with better ductility than conventional shearwalls, and wanting to show its clients that it was developing innovative technologies more quickly than its competitors, Obayashi's R&D division began research into higher-ductility shearwalls. Through its regular meetings with representatives of the design and construction divisions, practical parameters for efficient constructability were determined, and the decision to pursue the precast-panel, cast-in-place-frame concept was made. A team consisting of about six R&D employees, led by Dr. Kenzo Yoshioka, was formed, and began using computer models and full-scale mockups to study the structural performance of various shearwall configurations. By 1984, the team was ready to work with the firm's design and construction divisions to apply the system at an actual twelve-story structure in Osaka.

Obayashi is currently working on refinements of its shearwall system in anticipation of applying for approval of its use in tall structures. Use of the system on taller buildings, particularly those taller than 45 meters, requires approval by the Building Center of Japan (BCJ), an independently funded organization that works closely with, and under the jurisdiction of the Ministry of Construction. The BCJ puts much emphasis on full-scale tests, which explains in part why each comprehensive has invested in large structural testing facilities.

During the two years between the onset of the project in 1982 and the construction of a prototype hybrid shearwall in a commercial building, the input of both the firm's design and construction divisions was sought at certain milestones, to ensure the systems practicality and buildability. The ongoing research is still being done under the control of the R&D division, but representatives of the firm's structural engineering and architectural divisions meet with the R&D staff periodically to discuss the project. These interdivision interactions are fairly typical among the comprehensives, and are perceived as helping R&D projects find practical application while educating the production staff about emerging technologies developed by their R&D colleagues.

Obayashi has patented the basic hybrid shearwall system. In spite of the patent, Obayashi reports that similar systems have since been developed by its competitors through legal loopholes. Such replication is commonplace among the comprehensives and is one reason their major R&D efforts seem so similar. All the comprehensives have comparable core R&D capabilities, keep abreast of what their competitors are up to and have a keen understanding of the loopholes in Japan's patent laws. As a result, the developer of a significant advance in a core technology can expect its competitors to rapidly follow suit. The implementation of base isolation systems is a good case in point. Japan's first base isolation system was designed by the Tokyo Kenchiku Kenkyusha Co. for installation in the Christian Museum commemorating Miki Sawada, built in 1985. Obayashi was the first comprehensive constructor to install a base isolation system; its system was first constructed in 1986 at its Facility for Technical Research, located outside Tokyo. Just two years later, each of the Big Six comprehensives could boast the successful application of a base-isolation system in a full-scale building.

The comprehensives are also currently working to create electronic building databases, which would contain almost all the information needed to analyze and design a structure's components, estimate its cost and view its interior via an animated tour. Taisei, for example, is currently embarked on a six-year, $8 million project to develop an integrated database system called "Loran-T."[15] The system is planned around two databases. The first is a building database that would include the specifications for virtually all of a building's primary elements, including structure, HVAC, envelope, partitions and lighting. The second is a components database, which would store information on most of the parts the building could be made of. For example, the geometric and material properties of a particular steel beam on the fifth floor of a proposed building would be stored as part of the building database; geometric and materials properties for all steel sections available in Japan would be stored in the components database. As the building's design progressed, the building database would be extended and refined by all the designers working on it. Using HVAC design programs that draw on a projected building's volumetric data and facade system, for example, mechanical engineers would design the building's environmental control system. Subsystems and ducts would be drawn from the components database; chosen components would be automatically added to the building database. Of course design-architects would also use the system, which would be especially helpful for client presentations. Drawing from the building database, architects could simulate the spaces inside the building and quickly show a client the effect of a change in finish material or color.

The Loran-T program is not without potential problems and does not represent a large technological leap. The proposal, which emphasizes the use of predefined components, could reduce the creation of architecture to a sophisticated building-block exercise. Similarly, it would not incorporate any significant new computer or building design technologies. All of its systems would consist of current, state-of-the-art technologies. Even the ability to create animated tours through the building is now available on mainframe computers, and this technology is quickly being developed for PC computers. The Loran-T's most interesting aspect is its potential to integrate information traditionally stored in many independent sources. By storing all of a building's technical data in one place, Taisei would make it easier for everyone to find the information they need, while reducing mistakes made in transmitting information. Constructing a similar building database system in the U.S. would be more difficult, because of the number of independent consultants working on a typical job. For example, the information on a building's HVAC systems is typically held in the U.S. by a mechanical consultant, who transfers information on the weight and size of the system's components to the building's architect and structural engineer by fax or drawing. To change this system to an integrated, electronic approach would require everyone involved in a building's design and construction to agree to use compatible software, and indeed to use software for the entire design of a building, rather than relying on a mix of electronic and manual means as is now common. With their broad, in-house capabilities, the comprehensives can integrate existing technologies with relative ease.

The comprehensives also use R&D to help them diversify. Taisei's recent development of a concrete admixture called "Biocrete," illustrates their efforts to expand into construction-related product marketing. Taisei created Biocrete in a joint venture with Takeda Chemical Corporation, of Tokyo. The product improves the flowability and homogeneity of concrete, which reduces the need for concrete vibrating, and therefore saves construction labor costs. Biocrete is not the product of a traditional chemical lab; a bacterium grows some of the admixture's ingredients. Developed in two years as a collaborative effort between Takeda's chemists and Taisei's structural and bio-technology R&D staffs, Taisei is using the product in its own design-build contracts. Takeda and Taisei are currently discussing a marketing agreement in anticipation of selling the product to the concrete industry.

As explained in the discussion of Fujita's Automated Vertical Transport System (AVTS – Chapter 2), construction automation in Japan has focused primarily on roboticizing tedious or dangerous jobs, rather than increasing productivity itself. In general, construction automation systems are being developed slowly in Japan, and are being used to automate only the simplest construction tasks. The development of automated machines for building inspection and maintenance has produced better results. The Mitsubishi Electric Company, for example, developed a facade-cleaning robot in the 1970s that has since found many applications on Japanese facades, including Fujita's headquarters building. Similarly, clean room inspection robots have been developed by Takenaka, Kajima, Obayashi and Taisei. Takenaka's and Obayashi's systems, which are arguably the most advanced, may soon be marketed in the West (figures 11 and 12).

One reason for the success of the automated building maintenance and inspection systems is that the jobs they are asked to perform, like the work of their cousins in the auto industry, are simpler and more repetitive than the complex, unpredictable, often-similar-but-never-identical tasks that comprise most construction activity. The comprehensives have used this fact as the basis for developing their most striking automated building construction research, which has no analog in the U.S. Unlike roboticizing traditional construction tasks (or maintenance operations), automated building construction systems set out to transform the construction site into a place of automated assembly. The proposed process would be similar to a large manufacturing plant, whose product is the extension of itself. Robots inside this plant would fabricate a building's primary technical systems, including structure, building envelope and environmental control, from premanufactured pieces. Obayashi's "ABCS" (Automated Building Construction System), which is among the most advanced in the industry, illustrates the idea (figure 13). In this scheme, an automated all-weather factory would sit on jacks on top of a building's main columns, performing all the work required to assemble the structure, facade, and environmental control systems for the floor directly below it. After a floor is completed, the factory would jack itself up one floor and repeat the process (figures 14, 15, and 16). Obayashi has tested portions of this system, including automated placement of structural steel and jacking of the factory, on its Sumida Bachelor Dormitory in Tokyo in 1993 (figure 17). The company expects to expand the use of this system in constructing Arata Isozaki's Kashii Twin Towers in Fukuoka City, Kyushu (figure 18).

The Shimizu corporation tested portions of a similar system (called Shimizu Manufacturing System by Advanced Robotics Technology, or SMART) on its Nagoya Juroku Ginko Building in Nagoya in 1991 and 1992. The Nagoya Juroku prototype featured a self jacking weather enclosed factory, which was used for automated placement and welding of the building's steel skeleton, and for remote-controlled placement of some facade and floor panels.[16] In 1993, Taisei also tested a self-jacking, semi-enclosed platform for the erection of a building's structure. Called T-UP, Taisei's apparatus was used for erecting the steel skeleton frame of the Yokohama Building of Mitsubishi Heavy Industries (figure 19). Steel erection was performed by cranes mounted on the factory floor; welding of connections was performed manually.

11

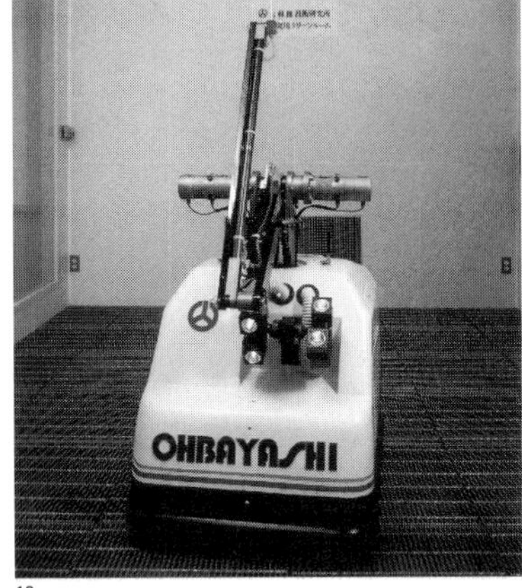

12

11. Takenaka clean room inspection robot.

12. Obayashi clean room inspection robot.

13. Obayashi's Automated Building Construction System.

13

14

15

16

14. Column positioning.

15. Column welding.

16. Facade panel installation.

An automated factory sits on jacks on top of alternate columns. Jacks positioned over the remaining columns are retracted they will be extended when another column-lift is positioned below them and construction of the floor below has been completed. To begin work on a given floor, column sections are delivered by a conveyor system, and positioned below the retracted jacks by an assembly robot. A welding robot then welds the column to the section below it (figure 15). Floorbeams are then placed by a ceiling crane and assembly robot. Next, prefabricated facade panels are lifted into place an robotically secured (figure 16) while a floor-laying robot installs the floor and mechanical equipment an utilities are installed between the floorbeams. Finally, the jacks positione over the columns just installed are extended while the adjacent ones are retracted, moving the automated factory upward one floor. The process is then repeated in the assembly of the next floor.

17

18

17. Prototype of portions of Obayashi's Automated Building Construction System (ABCS). Riverside Sumida Bachelor Dormitory, Tokyo, 1993.

18. Kashii Twin Towers, Model Photo. Arata Isozaki, Fukuoka City, Kyushu, Japan. (Photo by Yoshio Takase)

19. Erection of Taisei's Yokohama Building of Mitsubishi Heavy Industries using the "T-UP" System.

19

A complete automated factory system would at a stroke solve many problems plaguing Japan's construction industry today and would resolve many difficulties the Japanese have encountered trying to automate construction piecemeal. By erecting the structural, cladding, and most of the HVAC systems with robots, the need for construction workers would be significantly reduced, helping to alleviate Japan's labor shortage. The savings from reduced labor (depending on the cost of developing and maintaining the required robots) could be enormous. Productivity would also be boosted by incorporating all major assembly operations in the weather-protected roof-shed, and by the system's ability to easily function around the clock. Shimizu estimates that when completed, its system will reduce on-site construction labor and erection time by 50 percent. The work of automated machines, as the example of the auto industry shows, also produces consistently higher levels of quality than found in manual construction. Most important, because the system would transform the construction site into a manufacturing plant, jacking itself upward a floor at a time, the robots working in the plant would not need to be very mobile. This greatly simplifies their design and makes the system conceivable with technologies similar to those Japan already uses in its auto industry.

The system also has some drawbacks. Because it does not envision the inclusion of a heavy crane, the system could not install chillers and other heavy or bulky equipment without their first being broken into pieces. As these pieces are unique and complex, they would have to be assembled by hand at the site, reducing some efficiency gained by automation. A bigger problem is that the automated assembly machines would have limited mobility. This would require that the building plan, and to a lesser degree its section, would need to be regular. Indeed, Obayashi's proposed system envisions a circular or rectangular-plan building, and Shimizu's and Taisei's prototypes were based on a square building plan. This severely limits the application of the system for general use.

Technology Transfer

Since the onset of the Meiji era, Japan has been extraordinarily successful at assimilating and improving upon Western technologies. Today, the Japanese construction industry carefully monitors emerging trends in Western research and practice, and brings new product-development research ideas to Japan all the time. The frequent authorship of the comprehensives' researchers in prestigious Western engineering journals demonstrates that they keep abreast of the latest R&D efforts. In addition to keeping up with Western journals, the comprehensives' representatives visit foreign consultants, manufacturers, and universities on their own and through industry groups, to see what is emerging and how things are done. For example, Japan's Technology Transfer Institute, a consulting firm with branch offices in New York, Los Angeles and Paris, organizes information-gathering junkets to the West for the comprehensives' representatives, who come to observe and report on current methods in building design and construction practice.

One example of Japanese appropriation and adaptation of Western building technology is the development of the Tokyo Dome, or "Big Egg." The cable-stiffened, air-supported dome, which was designed and built by Takenaka in 1988, is based on similar roofs designed by the well-known American engineer, David Geiger. Geiger designed the first dome of this type for the 1970 Expo in Osaka. Japanese government approval for this structure, based on analytical studies by Geiger and smaller buildings previously erected in the U.S., was given on the condition that the structure would be temporary. In 1981, Takenaka's R&D division began working with Geiger toward the approval of permanent inflated cable domes in Japan. The government required that Takenaka perform extensive analytical and full-scale-mockup tests. In 1982, after performing various computer analyses, Takenaka built a 25 meter by 25 meter prototype roof, to study strength, fire prevention and maintenance aspects of the design (figures 20 and 21). Concurrently, they performed wind-tunnel tests (using their own wind tunnel) to predict the loads on the structure during typhoons. Based on the resulting data, a design manual for these roofs was compiled and submitted to the Building Center of Japan, which approved the design for medium spans. In 1985, based on further large-scale tests and the success of an inflated roof it built for a Buddhist group, Takenaka received approval for the 201 meter by 201 meter Tokyo Dome, which was completed in 1988.

Summary: R&D in Japan and the U.S.

Many of the comprehensives' R&D projects are not groundbreaking, and are analogous to work being done in the West. Geiger's inflated roofs and more recent developments by Geiger and Matthys Levy[17] illustrate that new building systems are being developed in the U.S., although not in construction firms and not at the rate at which they are emerging in Japan. Similarly, new building materials marketed by companies as disparate as W. R. Grace (glued structural connections) and Georgia-Pacific (microlaminated wood beams), show that American manufacturing firms are creating new products in their laboratories. Though it is hard to imagine the American construction industry doing much R&D work in building amenities, these areas are being pursued at universities and by some manufacturers. The U.S. can also claim prominence in building vibration R&D; much of the basic research on the effect of vibrations on human comfort levels has been done in the U.S., and manufacturing firms like 3M are developing vibration isolation systems in their labs. Finally, intelligent building systems are also being developed by American computer and HVAC companies.

20

20. Fire test of prototype air supported roof.

21. Load test of roof.

21

The main differences between the two countries' building design and construction R&D efforts are that significantly more R&D is being done in Japan, and that it is performed in different types of institutions in the two countries. Although the types of research done in each country are generally similar, there are some important fields of research, that if not uniquely Japanese, are not being aggressively pursued in the U.S. Most prominent among these areas is automated building construction, as exemplified in Obayashi's, Shimizu's and Taisei's systems. Because the scale of this research is so large, and because it involves virtually every discipline involved in building design and construction, it stands out as something that would be difficult to duplicate in the West.

Even given its limitations, automated building construction would be a revolutionary technology with the potential to produce awesome results. Its development says much about the forward-looking technical prowess of the comprehensives; its lack of development in the U.S. also underscores the highly specialized, fragmented organization of our industry, and its inability to envision systems and methods embracing a variety of professional disciplines and construction trades.

Table I: R&D Capabilities of Japan's Largest Comprehensive Construction Companies

Firm	Sales Volume (1991) (Million Yen)	R&D Budget (% of Sales)	Full-Time Staff	Year Founded
Shimizu	1,883,480	0.85	400	1946
Kajima	1,701,661	1.13	417	1949
Taisei	1,548,878	0.81	375	1963
Takenaka	1,403,075	1.01	269	1959
Obayashi	1,331,809	0.96	335	1948
Kumagai Gumi	1,201,413	0.38	94	1962
Fujita	734,718	0.62	140	1950
Toda	730,492	0.48	87	1958
Hazama	687,092	0.59	164	1945
Nishimatsu	552,059	0.66	174	1950
Tokyu	521,007	0.33	62	1971
Haseko	520,972	0.22	28	1987
Sato	520,873	0.42	80	1972
Mitsui	502,187	0.52	98	1975
Maeda	474,955	0.49	90	1973
Penta	439,061	0.55	91	1967
Tobishima	419,562	0.27	116	1967
Konoike	377,428	*	62	1963
Sumimoto	364,719	0.36	60	1960
JDC	343,618	0.44	90	1960
Okumura	340,088	0.46	77	1964
Aoki	323,593	0.55	43	1959
Zenitaka	298,937	0.21	35	1970
Tokai	286,256	*	25	1968
Asanuma	248,635	0.16	*	1987
Dai Nippon	243,850	0.14	27	1987
Ando	238,777	0.32	24	1988

* Not publicly available

Sources: *Japan Company Handbook 1992,* Toyo Keizai, Inc., Tokyo.
Takenaka Corporation, Tokyo.
Building Contractors Society of Japan, Tokyo.

Table II: Example Comprehensive Construction Company Research Areas

Subject	Prototypes Tested?	Used in Full-Scale Structures?
Intelligent Office Systems	Yes	Yes
Individually Adjustable Environmental Controls	Yes	Yes
Comfort Adjustment By Air Velocity Control	Yes	No
Olfactory Control Systems	Yes	Yes
Tensegrity Structures	Yes	No
Automated Warehouses	Yes	Yes
Automated Construction	Yes	Yes
Nonstructural Vibration Control	Yes	Yes
Structural Materials	Yes	Yes
Nonstructural Materials	Yes	Yes
Soil-Structure Interaction Dynamic Computer Modelling	Yes	Yes
Automated Construction	Yes	Yes
Design-Manufacture-Construction Integration	Yes	No
Biological Water Filtration	Yes	Yes
Grey Water Recycling	Yes	Yes
Clean Room Design	Yes	Yes

Source: Building Contractors Society, Tokyo.

Chapter 3 Endnotes

1. The R&D capabilties of the "Big Six" are described in detail in: Sidney Levy, *Japan's Big Six* (New York: McGraw Hill, 1993).
2. B. Paulson, *JTEC Panel Report,* 20.
3. Civil Engineering Research Foundation, *Transferring Research Into Practice,* chap. 4.
4. Hasegawa, *Built By Japan*, 160.
5. S. Levy, *Japanese Construction: An American Perspective,* 342.
6. C. Prestowitz, Jr., *Trading Places* (New York: Basic Books, 1988), 282.
7. "Used Cars in Japan," *The Economist* (December 21, 1991), 85.
8. Author's interview with Yoshida, June 1991.
9. Paulson, *JTEC Panel Report,* 25.
10. Like the fire control scheme, the schematic structural documents for Century Tower were prepared by one of Foster's long-time technical collaborators (Ove Arup and Partners in this case). Obayashi performed the final structural design and prepared all the construction documents for all the building systems.
11. Halpin, *JTEC Panel Report,* 76.
12. Author's interview with atmospheric researchers at Taisei's Technical Institute, 28 June 1991.
13. S. Suzuki, N. Kageyama, K. Nohata et al., *Active Vibration Control for High-Rise Buildings Using Dynamic Vibration Absorber Driven by Servo Motor* (Obayashi Corporation Internal Report, undated).
14. As noted in Chapter 2, the ductility of a structural component is a measure of how much energy it absorbs before it collapses (its strength is measured by the load that causes it to break).
15. P. Gillin, "Mixing High Tech and High Rises," *Computerworld* (August 13, 1990), 25.
16. Y. Miyatake and R. Kangari, "Experiencing Computer Integrated Construction," *Journal of Construction Engineering and Management* (June, 1993), 311-315.
17. M. Levy, "Floating Fabric Over Georgia Dome," *Civil Engineering* (November 1991).

Chapter 4

The Role of Government and Higher Education

The Ministry of Construction: Catalyst and Client

To modernize the country, Japan's government began creating new industries and transforming existing ones at the onset of the Meiji era in 1868. The ministries that currently control Japan's building design and construction industry stem from this tradition and can be traced to Japan's industrial reorganization after World War II.[1] Both the Ministry of International Trade and Industry (MITI, founded in 1949), and the Ministry of Construction (MOC, founded the same year) exert considerable influence on Japan's construction industry. MITI, familiar to Westerners as a force behind Japan's rise to preeminence in the automobile and electronics industries, also supervises aspects of the construction industry. As the agency primarily in charge of regulating both Japan's international industrial growth and its major manufacturing industries, MITI regulates the steel industry, and has an influence on housing, international construction and construction automation. It is MOC, however, that is mostly responsible for supervising the building construction industry. MOC maintains Japan's national building code, licenses the country's 2,900 specialized license contractors,[2] helps guide the construction industry's R&D efforts, and performs R&D itself at its Building and Public Works Research Institutes. MOC also publishes an annual white paper on Japan's construction industry, outlining the state of the industry and its priorities for research and growth.[3] MOC's importance is underscored by the fact that Japan's construction industry represents 20 percent of the gross national product (GNP), compared to about 9 percent in the U.S.[4]

The relationship between MOC and the industry it regulates differs from MITI's relationship to the auto and electronics industries in a few significant ways. Most obviously, MOC is not just an industry regulator but is also one of the construction industry's largest clients. MOC, or agencies working closely with it, commission most of Japan's large civil engineering works. As the comprehensives generally have large civil engineering departments, and because MOC decides their licensing status, the ministry's role as a client is significant. Also, unlike MITI, MOC tries to advance the technological capabilities of many firms, rather than consolidate advances (and market share) among a few companies. The market share of the Big Six comprehensives, at about 10 percent, is well below that of Japan's main auto and electronics firms, and MOC takes a laissez-faire attitude toward this. Finally, by long-standing tradition, the head of MOC is usually an engineer (MITI's top minister is almost always a lawyer).[5]

MOC's initiatives for building design and construction R&D originate with its Building Research Institute (BRI), located in Tsukuba. The BRI initiates and administers R&D programs for itself to carry out, in addition to its collaborative efforts with universities and industry. Although the American Society of Civil Engineers founded the Civil Engineering Research Foundation in 1989, in part to respond to MOC's initiatives,[6] its work is largely restricted to government-funded infrastructure research projects.[7] There is no American equivalent to the BRI.

With a staff of 171 researchers, administrators and support personnel,[8] the BRI is about half the size of a Big Six R&D division (see Table I, Chapter 3). Its 1991 budget of 2,470 million yen is significantly less than the Big Six's average 13,000 yen. What the BRI lacks in staff and budget, it makes up for in facilities. Its sprawling Tsukuba campus boasts seismology, geotechnical, structural, construction, wind and rain, materials, fire testing and irradiation labs. BRI also has the largest structural testing reaction wall in the world, which it installed in 1978 (figure 1). The most ambitious research to-date that used the reaction wall was the static and dynamic testing of a five-story, full-scale masonry structure. This unprecedented work was performed in 1987, as part of a joint research project with the U.S.[9] The BRI welcomed American participation not only to further inter-country cooperation, but also because it could not have executed the project using only its own budget.

BRI's "Roboticized Wall Erection Program" exemplifies its self-initiated and largely self-executed research projects. The program was begun in 1983 as part of MOC's "Development of Advanced Construction Technology by Applying the Electronic Technology" initiative.[10] According to Dr. Shuitsu Yusa of the BRI, the purpose of the program was to explore how it could adopt existing robotic technology, initially developed for the auto and electronics industries, to building construction. Working with university consultants, the BRI researchers developed a modular system of concrete masonry blocks and reinforcing, which a preprogrammed, essentially off-the-shelf robot assembled into rectilinear wall systems (figures 2 and 3). After the robot completed the assembly of a one-story wall system, the walls were manually grouted from above and precast planks covered the finished space. By 1987, the BRI had a working prototype system, which built a couple of one-story experimental structures for the agency. The system has not found commercial application for several reasons: the masonry blocks, which are laid tight, must be manufactured with a very high degree of accuracy, which drives up their cost; openings in a building had to be a full story high; a completely mobile version of the robot was never built; and the

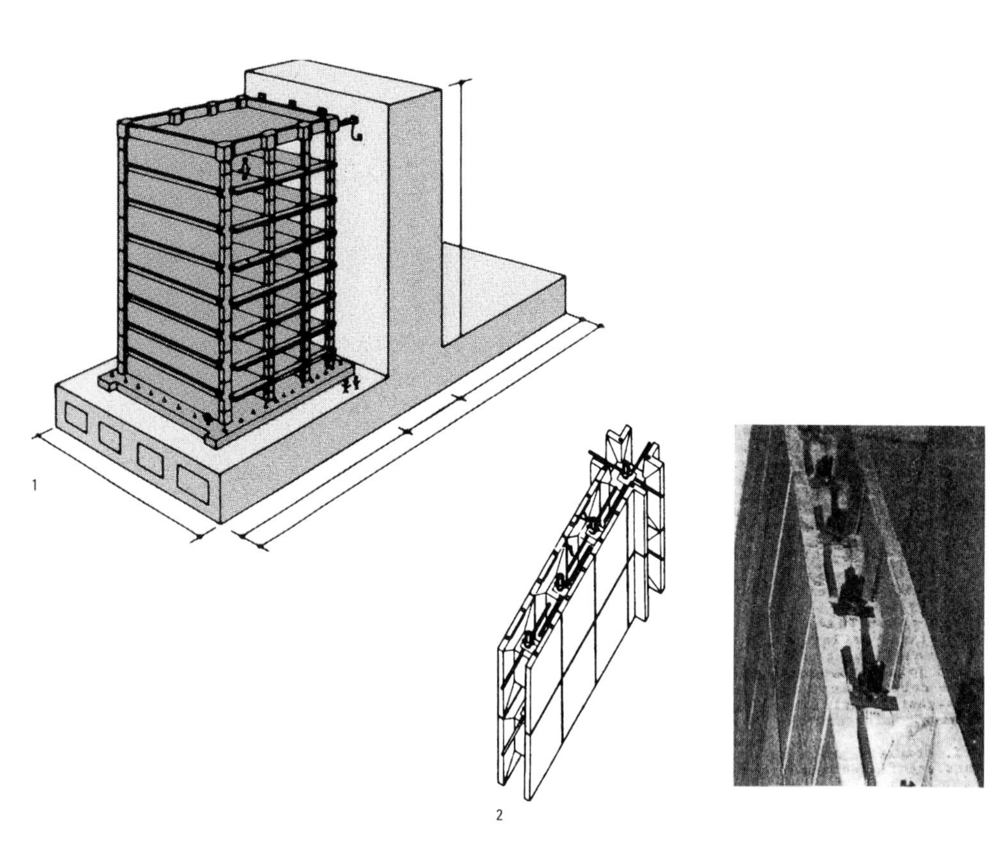

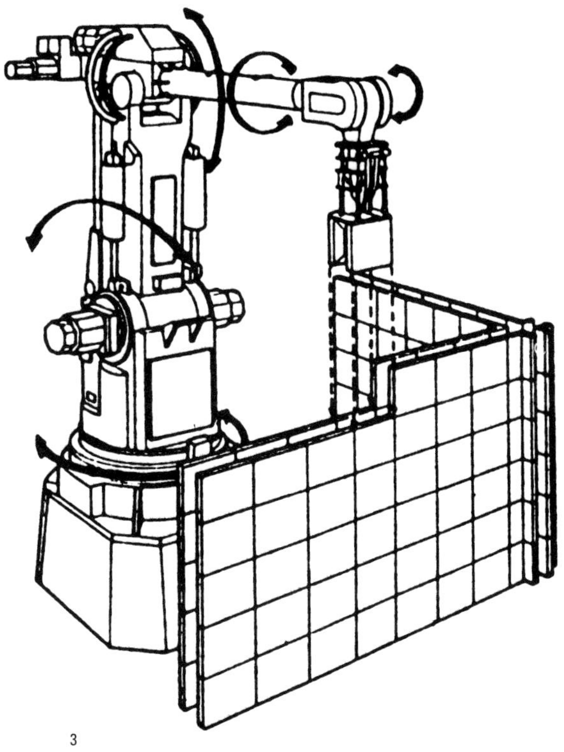

1. BRI's Structural Testing Lab, Axonometric View.

2. Roboticized Wall Erection System – Structural Components.

3. Roboticized Wall Erection System – Assembly Process.

robot worked slowly, so its contribution to improving productivity was negligible. The system's lack of real-world application did not bother the BRI, which was interested in gaining knowledge about automated construction rather than creating a commercial success.

A much more extensive campaign for design and construction automation is currently being waged by the BRI in one of its typical joint ventures with the construction industry. Each year, the BRI announces a series of research initiatives (typically intended to span about five years) and invites construction industry firms to participate.[11] The agency often also invites universities to participate in the research. The BRI contacts the comprehensives through the Building Contractor's Society, an industry trade association. In 1990 the BRI included as one of its research initiatives a program for "Advanced Construction Technology R&D," and invited the construction industry's participation. By 1991, twenty construction firms had agreed to participate and allocated R&D staff and funds. The goal of the five-year program is to develop an "Advanced Construction System," featuring automated design and erection of prefabricated reinforced concrete structures linking computer aided design (CAD), computer aided manufacture (CAM) and robotic construction technologies.

The system envisions designers working with a CAD system to lay out a set of predefined wall, floor and roof elements (figure 4). Once laid out, the CAD data would serve as the basis for the manufacture of precast components, which would be manually delivered to a staging area at the construction site. Installation would include both manual and robotic work, with robots placing the large reinforced concrete elements and humans fastening them into position.

The Advanced Construction System would only solve a building's structural and enclosure problems; workers would install HVAC (Heating, Ventilating, and Air-Conditioning), partitions, and finishes manually. Even with this restricted scope, the project is much more ambitious than the Roboticized Wall Erection project, which may help to explain why, as of June 1993, the BRI has not officially reported any preliminary results.

It is interesting to compare this system with the Automated Building Construction System (ABCS) proposed by Obayashi (Chapter 3, figures 13 through 17). Because it would not include the mechanical control systems planned by Obayashi, and because it would require more manual labor for assembling structure and enclosure than Obayashi's proposed steel-structure system, the BRI's proposal seems at first glance less progressive. But the BRI's idea is very forward-looking in two ways: first, it takes a comprehensive view of design, manufacture and assembly, and would integrate the three phases of building; second, like the blocks of its Roboticized Wall program, BRI envisions the assembly of the Advanced Construction System's elements in a variety of configurations, giving the designer a flexibility in space creation that is absent from the Obayashi system.

On balance, it is hard to say which of the two concepts is the most forward looking; by extension, it is hard to decide if the BRI's R&D programs lead or follow the comprehensives' individual efforts. In describing the Advanced Construction System, Akio Baba, head of BRI's construction techniques division, explains that one reason the BRI's effort envisions reinforced concrete structural systems is that the comprehensives are already developing automated construction systems using steel frameworks.

The ability of the comprehensives to pursue their own R&D initiatives explains, in part, their ambivalence toward participating in the BRI programs. In the words of Toshio Uetama, the Director of MOC's Foreign Affairs Bureau, "the General Contractors dislike us at MOC."[12] This may be overstated, but there are compelling reasons for the largest of the comprehensives to be wary of collaborating on R&D efforts with MOC. BRI designs its research projects to include not just Japan's largest firms but also smaller Zenecons. They also want the results of their work to be used industry-wide, not just by a particular project's participants. This is of course good news for the smaller Zenecons, which augment their R&D capabilities by working with MOC and the largest comprehensives. Even smaller firms with no R&D capabilities that would like to take advantage of technologies developed elsewhere, appreciate BRI's research philosophy. But to the comprehensives, which stand to lose their technological edge by sharing their work with companies unable to innovate on their own, BRI's position is not helpful.

The reasons the comprehensives participate at all in MOC's R&D efforts are complicated and to some extent vary by company. Since the comprehensives rely on MOC as a client for civil engineering commissions, it is in their interest to cooperate with them. Also, as MITI's experience in promoting the development of VLSI (very large-scale integrated circuit) computer-chip technology shows, to the extent that basic research is being done, spreading the work around helps everyone.[13] The comprehensives are also drawn to MOC's R&D as a way to promote new technologies they have already paid to develop. If a comprehensive's patented technology is incorporated into a MOC project and then becomes an industry standard, the company stands to gain through royalties. Regardless of who gains most by the government's R&D initiatives, the industry as a whole may well increase its competitiveness because of MOC's efforts. This is particularly true in comparison with the U.S., which is not vigorously pursuing similar research.

The University: Consultant and Talent Supplier

Both MOC and the comprehensives draw their employees from Japan's best higher education institutions, such as Tokyo University and Waseda University. The educational system that feeds these organizations is only superficially similar to its Western, particularly American, counterparts. As in the U.S., Japan's higher education program is based on four-year B.S. and B.A. programs, and advanced studies leading to masters and doctoral degrees. But the organization of Japan's universities within this framework is very different from that of the U.S. Tokyo University best illustrates Japan's system.[14]

Tokyo University's Department of Architecture is part of the Faculty of Engineering, and includes all engineering and architectural disciplines associated with creating habitable space. Civil, electrical, and mechanical engineering all have their own departments, but the architecture department offers its own required undergraduate courses in these subjects for its students. Architecture department faculty members teach these courses, which focus on issues clearly related to building design and construction. The Departments of Civil, Electrical and Mechanical Engineering provide their students with a more general introduction to these subjects. The architecture department's twenty-member faculty includes professors with expertise in such diverse fields as structural dynamics, architectural design, seismology, architectural history, fire protection and environmental engineering. The faculty's emphasis is weighted heavily toward the curriculum of American civil engineering departments. Graduates who intend to focus professionally on any area of building architecture or engineering all receive a B.S. in architecture from Tokyo University. All will have completed the same core education, including one and a half years of liberal arts, followed by required courses in structural physics and basic architectural design. The only distinction between the university's undergraduate-trained architects and engineers is their choice of departmental electives, which comprise forty of about 134 points required for graduation. Students who wish to practice architectural design do not need to study for five years, as in the U.S. In spite of its shorter length, Tokyo University's B.S. program produces architects who are more technically proficient than their counterparts in the U.S. The architectural design abilities of these graduates, however, is not as strong as that of American trained practitioners. This is largely because the architecture department places so little emphasis (relatively speaking) on this area. Only three semesters of design studio are required of the department's undergraduates.

Students pursuing graduate studies continue to work within the architecture department, but specialize in areas such as structural, mechanical, or electrical engineering, or architectural design. All the students in the two-year masters program receive a Master of Architecture degree; those completing a Ph.D. course receive a Doctor of Engineering degree, regardless of their area of specialization.

Japan's professional licensing system is also very different from that of the U.S. After two years, all of the students trained in the architecture department are eligible to take a national exam to become licensed architects (regardless of what they have done since graduation). Those practicing in fields as disparate as structural engineering and architectural design may obtain this license. Ensuring that large numbers of their young recruits become licensed architects is important to the comprehensives because these companies' status as general contractors depends, in part, on their employing a cadre of registered practitioners (see Chapter 1). For most other building-related firms licensing is less important. The chairman and former president of Nikken Sekkei (Japan's largest architectural firm), for example, is unlicensed, as are some of Japan's most renowned contemporary architects; the lack of a license clearly has not adversely affected their careers or dampened the fortunes of their design firms.

Professor Koyama of Tokyo University's Department of Architecture, explains that the department's organization dates to the inception of the Meiji era in 1868, when it was thought to be the most efficient way to help Japan catch up to the West. By integrating all the disciplines required to build a building in the same department, the university maximized the practical knowledge given to its graduates. This in turn helped them accelerate Japan's massive building effort.

Today, the multidisciplinary nature of Japan's university system informs both the organization of the comprehensives' design teams and their emphasis on design-build contracts. Like the organization of Tokyo University's architecture faculty, the comprehensives' approach to design-build commissions ties their design and construction departments together, minimizing the distinctions between them (see Chapter 2). Like the national goals Japan set for its universities during the Meiji era, the comprehensives' methods foster close and continuing communication among disciplines, which they claim leads to more efficient job management. The comprehensives' design construction teams, like the curriculums of Japan's universities, are populated primarily by the technically inclined. These teams tend to design spaces that are quick to construct and easy to maintain, if architecturally uninspired.

Western corporate models have also influenced the comprehensives' organizational structures. Skidmore Owings and Merrill's ability to provide complete architecture and engineering services, and Bechtel's "engineer-constructor" methods have been embraced by the comprehensives,[15] but these paradigms were established after the Zenecons had committed themselves to a team-design approach and began receiving design-build commissions.

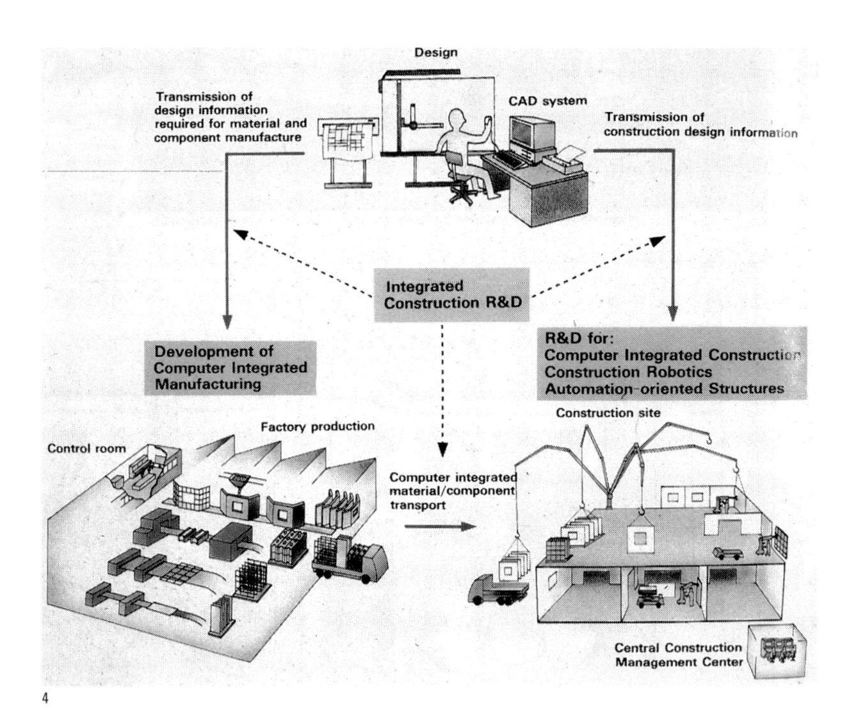

4

5

6

4. Schematic automated design and construction of precast concrete structure

5. National Gymnasium, Kenzo Tange. General view

6. Hisao Koyama, Main Hall for the Yokohama "Yes 89" Exposition, 1989.

Research Capabilities and Interaction with Industry

If the postwar period has marked the rise of the comprehensives' technical prowess, it is also distinguished by their eclipsing of university R&D programs in Japan. Today the R&D facilities of Japan's building design departments are relatively slim. The most prominent testing equipment used by Tokyo university's architecture department includes several midsize dynamic structural testing machines and a wind tunnel. All of the Big Six's R&D institutes have much more equipment, and unlike the architecture department, have full-time staff to run it. Professors at Japan's universities are also much less involved in actual building design and construction problems than they were thirty years ago.

The changes in the university's role in building design are epitomized in the contrast between the design and construction of Kenzo Tange's National Gymnasium, completed in 1964 (figure 5), and Professor Hisao Koyama's Main Hall for the Yokohama "Yes 89" Exposition, completed in 1989. Tange, realizing his cable-stiffened roof design posed some difficult structural problems, worked closely with Professor Yoshikazu Tsuboi of Tokyo University to develop the details. Professor Tsuboi, one of several engineers in Japan (or elsewhere) able to cope with this structure at the time, performed all the structural analysis and much of the structural design for the roof. The Shimizu Corporation built Tange's structure. Mr. Shunsuke Yamamoto, an executive with Shimizu, admits that Shimizu probably did not have the capability to perform the structural design of the National Gymnasium at the time it was built. He feels that today, however, Shimizu could easily provide structural design services for a similar building, and could also determine appropriate building envelope materials and environmental control systems. Professor Koyama agrees, and points out that he never works with his engineering associates at the university on equally complex projects today. His Main Hall for "Yes 89," for example, was engineered and constructed by the Fujita-Kogyo Corporation (figure 6).

There are many reasons for the eclipsing of the university's R&D operations and its waning participation in building design and construction – one of the most important is economic. Although the U.S. government contributes about 50 percent of America's total R&D funding, the Japanese government's contribution to its country's total is only about 25 percent.[16] Government funding for university-based R&D has also slightly declined in recent years. The modest size of Tokyo University's architectural R&D program reflects the small amount of government funding it currently receives.

The relatively low amount of R&D funding that finds its way to Japan's universities is to some extent explained by the country's history and current government philosophy. From the mid-nineteenth century until very recently, Japan has perceived itself as trying to catch up with the West. During this time, R&D efforts were not deemed necessary when methods of technology transfer were available, and R&D efforts that were undertaken were expected to produce practical results quickly. While the comprehensives have grown and prospered in the postwar period, they have continued to embrace these two ideas. Since World War II, the comprehensives have exploited the easily observed and openly reported nature of the construction industry to transfer Western technology to Japan; they have also closely linked their R&D divisions to their design-construction operations, and emphasized short-term R&D projects with clear applications in building design and construction (see Chapter 3). As James Abegglen has recently pointed out, this has suited the government, which has said that "as a matter of principle, technological innovation should be the job of the innovators – namely, businesses themselves."[17]

University professors participate less in building design and construction problems, in part, because they have become too specialized to be helpful in practice. Professors in Tokyo University's architecture department (and at most schools in the U.S.) list their areas of research in such narrow and arcane subjects as durability of concrete, seismic behavior of wooden buildings, and wind effects on structures. These researchers typically work primarily with one material, and sometimes with only one set of the material's physical properties (such as its thermal or structural performance). As design and construction problems become more complex and tend to overlap, they are less readily solved by these increasingly specialized professors. For example, the roof of Koyama's Main Hall building is made of steel, aluminum and Teflon-coated fiberglass fabric, and it performs structural, building enclosure and environmental control functions. Solving the roof's broad range of engineering problems requires greater breadth and teamwork than narrowly focused experts can easily provide. The comprehensives' team approach, on the other hand, is ideal for coping with just such complexities. This is reflected in their ability to handle almost any building design or construction problem with only their own staff.

Comparing university professors with colleagues within the construction industry, Dr. Hirsohi Akiyama of Tokyo University's Department of Architecture says "We like to think that although we have few researchers, we have a lot of brains." This is certainly true, and is one reason that university staff still sometimes contributes to the design of actual structures, and participates in joint R&D efforts with government and industry. For example, the work of the Robotics Research and Development Committee, which is researching construction automation, is partially funded by a government grant, and is headed by Professor Yukio Hasegawa of Waseda University.[18] Eleven companies are participating in the committee's work, including the Big Six, construction equipment manufacturers, electronics firms and a shipbuilding company. While industry representatives do most of the detailed development work, the university staff provides direction and expertise on difficult research problems.

As Japan's universities have relinquished much of their consulting and R&D roles to the comprehensives, these companies have mimicked the schools' organizational models and have drawn an increasingly sophisticated technical staff from them. In the process, the comprehensives have created efficient methods of building buildings and developed results-oriented R&D divisions. The transformation has created some obvious problems, however, especially for the schools. Even if they were so inclined, it would be hard for architecture departments to justify commercially applicable R&D at this time, because so much is already being done by industry. But as research shifts to the Zenecons, it also becomes more difficult for schools to fund their increasingly esoteric research projects, which do not have immediate market justification. The current system may also cause the comprehensives some grief. By populating themselves with narrowly trained professionals, they maintain a certain level of technical proficiency. Yet conceptual creativity in technical innovation and architectural spacemaking is often the product of a broader perspective. In the long term, the comprehensives' staffs may be best able to refine the advances of others, rather than pioneer their own innovations.

Chapter 4 Endnotes

1. C. Johnson, *MITI and the Japanese Miracle* (Palo Alto: Stanford University Press, 1982), 330; Prestowitz, *Trading Places*, 231.

2. As discussed in Chapter 1, contractors must be prequalified as specialized license contractors to be able to bid on many of Japan's largest infrastructure and public-sector design and construction commissions.

3. Ministry of Construction, *White Paper for Construction* (1991).

4. MOC, *White Paper for Construction*, 338.

5. Johnson, *MITI and the Japanese Miracle*, 60.

6. "Responses Must Be Immediate," *Engineering News Record* (September 13, 1990).

7. "Cerf Makes Plans for Highway Innovation Center," *American Society of Civil Engineering News* (November 1992).

8. MOC, *Building Research Institute*, Ministry of Construction, Japan (undated).

9. Building Research Institute, *Large Scale Structure Laboratory* (pamphlet, undated).

10. Y. Kodama et al., "A Roboticised Erection System with Solid Components," Proceedings of the Fifth International Symposium on Robotics in Construction (Tokyo, Japan, June 6–8, 1990).

11. Unless otherwise noted, information in this paragraph was obtained in author's interviews with BRI officials and members of Taisei's and Obayashi's R&D staffs in June 1991.

12. Author's interview with Toshio Uetama, June 1991.

13. As MITI's development of VLSI technology also demonstrates, this becomes less and less true the closer a given technology comes to prototype stage. MITI's experience showed that once basic research is completed, a company's first instinct is to forget collaboration in its race to deliver a marketable product. For details see: J. Abbeglen, *Kaisha The Japanese Corporation* (New York: Basic Books, 1985), 138.

14. Unless otherwise noted, the information in this section was provided by: author's interviews with Professor Hisao Koyama, The Department of Architecture, Tokyo University, Tokyo, 1991; University of Tokyo, *Information on the Faculty of Engineering, The Unversity of Tokyo* (1990); The Department of Architecture, University of Tokyo, *1991 Department of Architecture* (in Japanese).

15. Author's interviews with Taisei and Takenaka executives, Tokyo, June 1991.

16. Abbeglen, *Kaisha*, 137.

17. Economic Planning Agency, *Economic Survey of Japan* (1978), 137.

18. Bennett et al., *Capitol and Counties Report*, 61.

Chapter 5

Technology and the Competitive Edge

Benefits of Technological Advancement

The comprehensive construction companies' pervasive pursuit of technology directly benefits them in several ways. As the Taisei Corporation's development of "Biocrete" shows (see Chapter 3), the comprehensives' R&D divisions are sometimes used to help improve efficiency and diversify. Unlike many Western companies, whose diversification into areas they know nothing about has led to disastrous results,[1] the comprehensives have successfully expanded into areas related to building design and construction.

Biocrete's short development time (two years) is typical for the comprehensives, which like to find practical use for their research quickly. By contrast, some large Western companies, notably IBM, are equally as well known for their inability to bring technological advances to market as they are for the quality of their basic research.[2] While the comprehensives' short-term R&D projects may not produce many breakthrough technologies, they do turn out a steady stream of modest improvements to building design and construction. Over time, the cumulative effect of these incremental gains may be large.

The comprehensives have also benefitted by using technology to create demand for new building services. In Japan, where many corporate clients like their offices to be as technologically current as possible, the installation of high-tech amenities at one location has often instigated demand for similar systems elsewhere. Shimizu's Makuhari project (see Chapter 2), which features a variety of intelligent, automated tenant-service functions, was designed with this in mind. Some of the comprehensives have also tried to stimulate demand by using their corporate headquarters as technological showcases. Shimizu's headquarters and Kajima's KI Building, for example, are used by their sales staffs as full-scale demonstrations of many state-of-the-art systems.

During the 1980s, diversification and demand-creation were successful strategies for the comprehensives. In developing new technologies related to their primary businesses, such as automated building control and maintenance facilities, the comprehensives expanded in a relatively safe way. Until 1992, the Japanese reflexively used technology to improve the quality of life from the bathroom to the boardroom (figures 1 and 2). It was easy for the comprehensives to create markets for high-tech, high-cost amenities because corporate clients used these products to increase their cachet and societal approval.[3] The bursting of Japan's bubble economy, however, has reduced demand. The current frugal climate continues to make Japan's executives less eager to pay premiums for technological luxuries. In the rest of Asia and the West, where motion-sensitive light switches and auto-flushing urinals are just now becoming commonplace, interest in these amenities is less intense and unable to offset the weak market in Japan. Still, as Japan's economy recovers and as other countries become more familiar with high-tech gadgetry, demand for these products will grow and the comprehensives will be in a good position to provide them (figure 3).

In addition to their own pursuit of new technologies, the comprehensives' technological prowess benefits from their organizational structures, and from the organizational similarities between the comprehensives' and Japan's higher education institutions. By training architects and engineers in the same department, Japanese universities prepare them to move smoothly into the comprehensives' team structure, which maximizes interaction between building designers and constructors. Projects like Takenaka's active mass damping system (used at its Intes Building to enhance both the building's structural and mechanical performance) show that this strategy can produce creative solutions to technological problems.

The design process for the facades of Mitsui's River City buildings (Chapter 2) shows that the design-build system itself helps the comprehensives by allowing their in-house teams to make purchasing decisions at the last possible minute. This in turn encourages them to use the latest technology, besides helping them save money by postponing expenditures for building materials.

Many of the Japanese construction industry's technological advances have been criticized as refinements and adaptations of existing ideas, rather than original innovation.[4] Projects like the Intes Building notwithstanding, some of the comprehensives' R&D work underscores this, especially its emphasis on transferring building systems (such as tensegrity domes) from the West, and borrowing robotic technology from Japan's automotive industry. Similarly, the comprehensives were not the first to develop base isolation technology in Japan, which was brought into the country by a relatively small engineering company for use in the Miki Sawada Museum building in 1985. But some of the comprehensives' current R&D efforts represent innovative, original ideas that have great potential to transform the industry. Shimizu's, Obayashi's or MOC's construction automation systems would make it possible to transform the construction site from a place of building to a manufacturing plant – a potentially revolutionary event in the history of high-rise building construction.

1

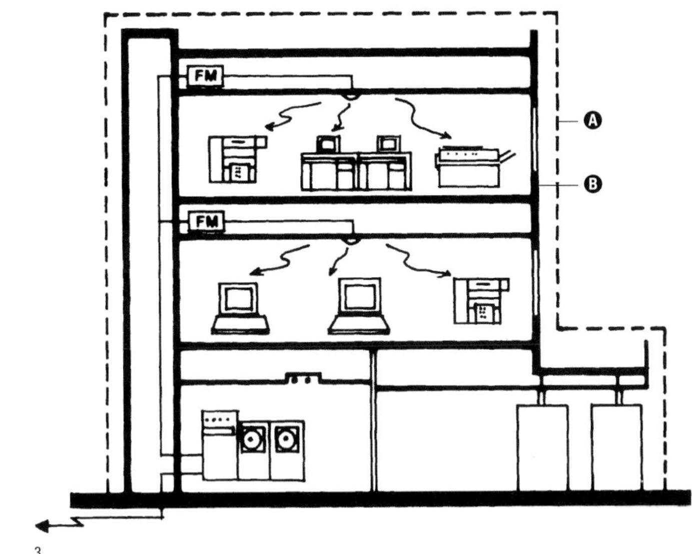

3

1. Bathroom emergency call button. Hotel Edmont, Tokyo.

2. Conference room featuring individual smoke exhaust systems. Onward Institute, Fujita Corporation.

3. Electromagnetic and physical building enclosures, schematic. Obayashi Corporation.
A. Electromagnetic envelope
B. Atmospheric envelope

2

The comprehensives' research efforts also gain significance through their continuity. Long-term support for R&D is encouraged in the *Keiretsu* system, which emphasizes increased market share over short term cash returns. In spite of the industry's retrenchment caused by the recent economic downturn, R&D continues to be important to the comprehensives, and it has not been financially gutted during the recent period of hardship. Their long-term support for research, along with their links to Japan's government and higher education system, have helped the comprehensives establish world-leading R&D divisions in an industry not known for generating technological advances from within. More important, the continued growth of the comprehensives' R&D divisions is helping to transform Japan's construction industry into a high-technology field.

A Model for the West?

This transformation, and the benefits the comprehensives reap from it, suggest that Western firms may want to emulate some of the Japanese industry's features. This idea should be approached with some caution, for a few reasons.

Many of Japan's methods may not travel well to the U.S. The structural differences in the two countries' industries, for example, would make it hard (and inappropriate) to transfer many of Japan's technological methods to America. Unions in the U.S., once champions of the common man, now often act together as a powerful, modern-day Luddite lobby, opposing new technologies that might decrease the need for manual labor. American construction unions could be expected to vociferously oppose any significant automation schemes. The comprehensives' corporate structures rely heavily on lifetime hiring policies and emphasize team productivity over individual stardom. America's entrepreneurial spirit, and its emphasis on individual merit, encourages employees to put their personal goals above their company's and discourages corporate loyalty. These attitudes would present substantive barriers to implementing the comprehensives' management techniques. The comprehensives' long-term outlook would also be hard to establish among American construction executives because American corporations are hindered by stockholders who are primarily interested in short-term gains. This makes executives less inclined to spend time and money developing new technology than their Japanese counterparts. American managers have also learned that technological innovation does not guarantee success in the marketplace (the disastrous experience of computer-chip makers in the 1980s provides a good example), and they have become reluctant to develop new ideas.[5] As a result, American firms cannot be expected to increase R&D outlays unless they become convinced – as the comprehensives already are – that such investment is necessary to ensure their survival.

Societal differences would also keep many Japanese methods from being easily transferred to the West. Japan's general faith in technology's ability to address many of society's problems fuels the comprehensives' technological explorations. Speaking of the construction labor shortage in Japan, for example, Obayashi's Sato Chikafusa expresses a commonly held view when he says "all we can do is automate" to solve the problem. Similarly, Japanese executives are more technologically sophisticated than most American managers. A glossy marketing brochure for Shimizu's Makuhari complex illustrates this. Amidst a splashy presentation describing the complex's healthy environment and the wealth of infrastructure supporting it, the brochure displays a tabulation of the allowable live loads on various sections of the complex. This information would be lost on most potential corporate tenants in the U.S., because their executives typically have little notion of what things weigh and don't know what a "live load" is.

The differing relationships between government and industry in Japan and the U.S. would also keep Japanese methods from translating easily. Japan's government *develops* rather than simply *regulates* its design and construction industries. As Chalmers Johnson points out in his classic *MITI and the Japanese Miracle*, "for more than 50 years the Japanese state has given priority to economic development,"[6] and technological advancement has been a prominent feature of the government's economic development strategy. This is reflected in the construction industry by MOC's R&D efforts, which it often performs in partnership with construction companies. In the U.S., by contrast, the federal government concentrates on ensuring that everyone plays by the rules, and, like the construction industry itself, makes few R&D initiatives. Except a few nascent efforts at Bechtel's R&D division, the Army Corps of Engineers' Construction Productivity Advancement Research Program, the American Society of Civil Engineers' Civil Engineering Research Foundation, and the independent Construction Industry Institute, construction R&D initiatives are rare in the U.S.; joint government-industry programs are virtually nonexistent.

In addition to societal differences, some features of Japan's industry are in fact less appealing sources of inspiration than they seem at first glance. Both the organizational structures and methods of the comprehensives have some significant problems. In this author's view, architectural design, for example, is paid scant attention in many of these firms. This is reflected in the comprehensives' frequently not charging clients (explicitly) for architectural services in design-build contracts. The comprehensives explain that not mentioning architectural design in contracts is more a reflection of Japan's business customs than an indication of perfunctory treatment of spatial design.[7] They point out that traditionally the Japanese have expected to pay only for "hardware" (the completed building), and assumed that "software" (design and analysis) would be included implicitly in contracts as overhead. Still, this system's inevitable emphasis on hardware tends to steer the industry toward technological innovation at the expense of spatial felicity. The emphasis of Japan's architectural educators on technical prowess reinforces this trend. Proposed new construction methods and building systems – such as Obayashi's and Shimizu's automated building construction systems, Taisei's building database project, and BRI's automated construction system – show how these factors push the comprehensives toward the commodification of building design.

There is some evidence that the advantages of hiring for life and seniority-based pay are beginning to be outweighed by the corporate inflexibility it creates, and the worker lethargy its security sometimes encourages. Some Japanese business people now expect that Japan's tradition of lifetime employment will not continue to be universally embraced.[8] Also, some of the country's most successful small companies are experimenting with merit-based pay instead of Japan's traditional seniority-based compensation. The Keyence Corporation, for example, a small but quickly growing Osaka-based maker of magnetic sensors, considers its corporate meritocracy a central reason for its success.[9]

Finally, the endemic collusion and too frequent corruption in Japan's construction industry underscore the dangers of close cooperation among competitors and between competitors and government. As Brian Woodal has recently reported, there is a "pervasive and institutionalized system of bid rigging [operating] in Japan's Construction market."[10] This can hardly have a beneficial effect on the price paid for the industry's services. Equally troubling, as Woodal goes on to report, is the fact that "of the 100 or so bribery scandals each year in Japan, over half involve construction firms." Recently, Japan's press has reported on large bribes given to the leaders of the Liberal Democratic Party by construction industry executives,[11] and these scandals have contributed to the collapse of the government and the splintering of the party. In this context, it is important to note that Japan's construction industry bribes are used primarily as a method of obtaining work, not as a way to gain acceptance of poor workmanship. High quality construction is as real a goal as continued technological progress in Japan. This is evinced in both the workmanship of large, recently completed state projects, and the ostracism companies risk if their work is perceived to be deficient. In the U.S., however, construction industry corruption and poor quality work are inextricably intertwined. Close links among competitors and between government and industry might encourage more problems in the U.S., and should be approached skeptically.

Despite all the problems of Japan's construction industry, it has positive aspects that should not be overlooked. Its technological strengths in particular, are real and increasing. Even as they are undergoing retrenchment during Japan's current economic slowdown, the comprehensives are continuing to develop new products.[12] As previously noted, the Japanese industry's best R&D efforts are also creative and significant, and the radical automation systems currently being explored promise potentially huge gains in productivity. This is especially important in light of the boost that productivity gains have given to other industries. Due largely to automation, the productivity of the auto and electronics industries has doubled over a recent ten-year period, while auto costs have remained flat and the price of calculators has dropped one thousandfold.[13] Large gains in building design and construction productivity could be expected to have similar effects on the cost of completing a new building. At a time when building cheaply is considered paramount, the ability of automation to lower a building's cost while bypassing union labor may well prove irresistible to American developers. If Japan's industry is able to deliver on the promise of its proposed automated systems, it might soon be manufacturing buildings in the U.S. – making the West's construction business "the next industry to bear the brunt of Japan's exportation of high-tech products."[14]

Unfortunately, the U.S. is not paying much attention to Japanese industry's efforts. As recently reported in the House of Representatives, "The U.S. is deficient in scanning the world's science and technology for potential commercial opportunities relative to what is done by its competitors, particularly Japan."[15] Our myopic view, as painfully discovered by the American auto industry in the 1980s, can have disastrous consequences. For this reason alone, the American construction industry should take notice of what the Japanese are doing and formulate a response.

Two prominent parts of American industry's response should be to improve both long-term and short-term technological advancement efforts. To meet the first goal, the industry should establish a strong, continuing commitment to R&D, represented by increased and secure funding levels. To meet the second goal, the industry should initiate a short-term technology transfer program, to import Japanese technologies and some ideas. In the past, Japan has relied on technology transfer to catch up to the West, while it has built up its R&D capabilities. The American construction industry should now use the same technique to ensure its own future position. Otherwise, it risks finding itself – like America's auto and electronics industries before it – unable to compete against a more efficient and technologically advanced Japanese juggernaut.

If American industry embarks on a strong program of technological advancement, it will secure its ability to compete harmoniously with Japan in the world's emerging markets; if it doesn't, it may become obsolete. The progeny of Japan's pioneering automated systems may be assembling buildings outside the country's borders in the near future. If the American construction industry does not anticipate this, it may soon find Japan's automated hardware working here, erecting buildings more economically than the U.S. can itself.

Chapter 5 Endnotes

1. Prestowitz, *Trading Places*, 335.
2. "IBM: Push R, Pull D," *The Economist*, vol. 321 (30 November 1991), 81–82.
3. Civil Engineering Research Foundation, *Transferring Research Into Practice.*
4. Bennett et al., *Capitol and Counties Report*, 53.
5. Prestowitz, *Trading Places*, 120–161.
6. Johnson, *MITI and the Japanese Miracle*, 305.
7. Building Contractors Society, letter to the author, December 15, 1993.
8. "The Other Bubble," *The Economist* (May 29, 1993), 73–74.
9. "Bright Sparks," *The Economist* (May 15, 1993).
10. "The Logic of Collusive Action," *Comparative Politics* (April 1993), 297–312.
11. "Bribery Case Hits the Miyazawa Faction," *Tokyo Business Today* (February 1992), 12.
12. "Construction Machinery Shifts Toward Labor-Saving User-Friendly Machines," *Japan 21st* (February 1992), 61–71.
13. Abbeglen, *Kaisha*, 61.
14. *Capitol and Counties Report*, 67.
15. House Subcommittee on Investigations and Oversight and the Subcommittee on Science, Research, and Technology of the Committee on Science and Technology, Japanese Technological Advances and Possible United States Responses, 98th Cong., 1st sess., 29–30 June 1983, 17. [Quoted by Abbeglen in *Kaisha*, 61].

Index

Abegglen, James 92
acoustic-modeling. See STRADIA
active mass dampers (AMDs) 25-28
Advanced Construction System. See Building Research Institute
air-sprung truck suspension 25
Akiyama, Hiroshi 92
American Society of Civil Engineers 84, 99, 109
Army Corps of Engineers 99
Asano, Sadayasu 57
Automated Building Construction System (ABCS). See Obayashi Corporation
Automated Construction 15, 42, 43, 58, 81, 86, 87, 100
Automatic Material Transport System (AMTS). See Fujita Corporation
Automatic Vertical Transport System (AVTS). See Fujita Corporation

Baba, Akio 87
Big Six 12-14, 55, 83, 84, 91
Biocrete. See Taisei Corporation
Boston Symphony Hall 60
British Art Museum 32
Building Center of Japan (BCJ) 69
Building Contractors Society 9, 16, 43, 52, 80, 81, 86
Building Research Institute (BRI) 84-87

Chikafusa, Sato 32, 54, 99
Ciba-Geigy 46
Citicorp Building 28
Civil Engineering Research Foundation 9, 16, 82, 84, 99, 102
climate control systems 28
Columbia University 9, 16, 35, 109
Comprehensives 14, 15, 17-20, 25, 28, 32-37, 95-101
Comprehensive Construction Company. See Comprehensives

computer aided design (CAD) 35, 86
computer aided manufacture (CAM) 86
concrete 20, 22, 28, 31, 42, 43, 45, 59, 63, 64
  Biocrete. See Taisei Corporation
  carbon fiber reinforced concrete 31
  high-strength concrete 59, 63, 64
  Mitsui Steel-Concrete (MSC).
    See Mitsui Corporation
  post-tensioning 20
  prestressing 20
  reinforced concrete (RC) 20, 22, 31, 66
  steel-reinforced concrete (SRC) 20

dynamic vibration absorber (DVA) 66, 67

earthquake 11, 19, 20-28, 32, 53, 59, 63, 66
Empire State Building 25
environmental control systems 28, 34, 49, 65, 72, 91

fire shutters, heat controlled 32
Fluor Daniel Corporation 12
Foster, Norman 22, 23, 60
  Century Tower 23, 62
Fujita Corporation 28, 29, 43, 54, 97
  Automatic Material Transport
    System (AMTS) 46-48
  Automatic Vertical Transport
    System 43, 46, 71
  Makuhari Syokugyou Center
    Project 43
Fujita-Kogyo Corporation 91

Geiger, David 77
Georgia-Pacific 77
Griffis, Bud 9, 35, 54
Grösser Musikvereinssal 60

Hasegawa, Yukio 92
Hazama Gumi 18, 19, 21, 28
  Mitsukoshi Department Store 18, 19, 21, 28
House of Representatives 101
Hyatt Regency Hotel 32
hybrid structural components 20

intelligent building systems 48, 49, 58, 59, 77
International Business Machines (IBM) 95
Isozaki, Arata 72, 75
    Kashii Twin Towers 72, 75

Johnson, Chalmers 94, 99, 102

Kahn, Louis 32
Kajima Corporation 25, 31, 33
    Edo-Tokyo Museum 25, 26, 27
    KI building 32, 33
    Manhattan hotel 35
    World Business Garden building 34, 35
Kashii Twin Towers. See Isozaki
Keiretsu 55, 98
Keyence Corporation 100
Kidder, Tracy 39
Koyama, Hisao 89-92, 94
    Main hall for the Yokohama
        "Yes 89" Exposition 90-92
Kumagai-Gumi Corporation 12

Levy, Matthys 77
local area network (LAN) 49
Loran-T. See Taisei Corporation

M.W. Kellogg 11
Maki, Fumio 12
mechatronics 47
Miki Sawada Museum 96
Ministry of Construction (MOC) 32, 58, 83, 84, 87, 88, 99
Ministry of International Trade and
    Industry (MITI) 58, 83, 84, 87, 99
Mitsubishi Electric Company 71
Mitsui Corporation 20, 22
    H Building 22
    Mitsui Steel-Concrete (MSC) 22
    River City 20, 22, 23, 96
Mitsukoshi Department Store.
    See Hazama Gumi

National Gymnasium. See Tange
Nikken Sekkei 89
Nippon Steel Corporation 31

Obayashi Corporation 13, 14, 22, 60, 67-70, 73, 82, 97
Automated Building Construction System (ABCS) 72-75, 86
Century Tower 60, 62, 63
One Day One Cycle 42, 45

Portman, John 32
Preston, J. R. 63

research and development (R&D) 11, 12, 55, 91
robots 35, 42-43, 48, 51, 58, 72-76, 84-87, 92, 96
robotic construction 42, 86
Robotics Research and Development Committee 92

Sawaguchi, Masahiko 46-48
shearwall 66, 68, 69
Shimizu Corporation 12, 16, 19, 49-51, 57, 72, 76, 91
Automated Building Construction System 86
Makuhari Techno-Garden Complex 49-51
Nagoya Juroku Ginko Building 72
Seavans Headquarters Building 35, 52
SMART 72, 76
Skidmore Owings and Merrill 89
SMART See Shimizu Corporation
Soul of a New Machine (The) 39
Specialized License Contractors 12, 83, 94
Starnet Structures 35
steel construction 20, 22
eccentric-braced frames 20, 22
moment frames 20, 22
STRADIA 60, 64

T-UP See Taisei Corporation
Taisei Corporation 59, 70-72, 95
Biocrete 71, 95
Loran-T 70, 71
T-UP 72
Yokohama Building of Mitsubishi Heavy Industries 72, 75

Takenaka Corporation 12, 13, 23, 24, 28-32, 80
Asahi Newspaper Annex 22-24, 36-39, 41
Intes Office Building 12, 13, 28-32, 49, 52, 96
Tange, Kenzo 12, 31, 90, 91
National Gymnasium 12, 90, 91
Tokyo Metropolitan Government
complex 31
Technology Transfer Institute 76
Tokyo Dome 77
Tokyo University 9, 16, 88-92
Department of Architecture 88
Department of Civil Engineering 88
Tsuboi, Yoshikazu 91
Tuned mass dampers 28, 53

variable force tendons 25
vibration control systems 25, 53
viscous dampers 25

W. R. Grace 77
Waseda University 88, 92

Yamamoto, Shunsuke 91
Yoshida, Jin-Ichi 53, 59, 82
Yoshioka, Kenzo 69
Yusa, Shuitsu 84

Zenecon 12, 14, 16, 20, 43, 58, 87, 89, 93

Anthony C. Webster is Director of Building Technologies and an assistant professor at Columbia University's Graduate School of Architecture, Planning and Preservation. Mr. Webster is also a consulting associate at Weidlinger Associates, a structural engineering firm in New York City.

Mr. Webster received a B.S. in Applied Science in Engineering from Rutgers University in 1980, where he graduated summa cum laude. In 1984, he received an M.S. in Civil Engineering from Columbia University, where he was the recipient of the university's Carleton Fellowship and the CRSI's Cameron Fellowship. In 1987, with the help of a tuition grant from Weidlinger Associates, he received a Professional Degree in Civil Engineering from Columbia. Between degrees, Mr. Webster worked in New York City as a structural engineer at Iffland Kavanagh Waterbury, P.C. and at Weidlinger Associates.

Mr. Webster's articles on Japan's building design and construction industry have been published in the *Journal of Urban Technology*, *Architectural Record* and the American Society of Civil Engineers' *Journal of Professional Issues in Engineering Education and Practice.* His technical articles on tuned mass dampers have appeared in *The Engineering Journal* and *Civil Engineering Magazine.* Mr. Webster's analysis of Santiago Calatrava's bridge designs appears in *Calatrava: Bridges*, (Artemis, 1993) and in *The Architectural Review* (November 1992). He is a co-editor of *Bridging the Gap: Re-Thinking the Relationship of Architect and Engineer* (Van Nostrand Reinhold, 1991).

Mr. Webster's design projects have been published in *The Engineering Journal*, *Civil Engineering*, and *On Making.* His current research includes the development of an augmented reality prototype for visualizing structures (with Steven Feiner and Ted Krueger), and an analysis of the relationships among culture, technology and new architectural forms in twentieth century America (with Sarah Ksiazek).